Design and Equipment for Restaurants and Foodservice: A Management View

Design and Equipment for Restaurants and Foodservice: A Management View

Editor

Kornelia Marshall

scitus academics

Design and Equipment for Restaurants and Foodservice: A Management View

Edited by **Kornelia Marshall**

ISBN: 978-1-68117-249-1
Library of Congress Control Number: 2016934792

SCITUS Academics LLC,
www.scitusacademics.com
Box No. 4766, 616 Corporate Way,
Suite 2, Valley Cottage,
NY 10989

Notice

Printed in United States of America on Acid Free Paper

Preface

The food service industry encompasses any establishment that serves food to people outside their home. This includes restaurants, carryout operations, cafeterias, university dining halls, catering and vending companies, hotels and inns, and rehab and retirement centers. The foodservice industry is divided into two segments. The commercial segment makes up 80 percent of the industry and includes operations in restaurants, catering and banquets, retail, stadium, airlines, and cruise ships. The non-commercial segment includes schools and universities, the military, health care, business and industry, and clubs. Food service is a large industry which deals with the preparation and service of food outside the home. Catering a wedding, establishing a restaurant, and running a cafeteria are all forms of this service. A number of goods and services fall under the umbrella of food service, such as companies which transport food and related products like kitchen equipment, silverware, and so forth. Restaurateurs, waiters, bussers, chefs, and dietitians are some examples of people who work in the food service industry, along with people like architects who design facilities where food is made and served, company representatives who travel the road selling products related to the service and sale of food, and consultants who help people coordinate events at which food will be served.This book entitledDesign and Equipment for Restaurants and Foodservice: A Management Viewoffers the most comprehensive and updated coverage of the latest equipment and design trends to help students acquire the knowledge they need to go into the industry. The book emphasizes on how to plan, design, and purchase equipment for a restaurant, or foodservice facility.

Table of Contents

1

CHAPTER

A REVIEW OF LITERATURE ON RESTAURANT MENUS: SPECIFYING THE MANAGERIAL ISSUES

Bahattin Ozdemir, Osman Caliskan

Gastronomy and Culinary Arts, Akdeniz University Tourism Faculty, Dumlupinar Bulvarı, Konyaaltı, Antalya 07058, Turkey

ABSTRACT

This conceptual study specifically aims at reviewing the critical managerial issues of menu, and demonstrating the conceptual structure of menu management. Based on the conceptual and empirical findings of menu literature, the major menu management issues are menu planning, menu pricing, menu designing, menu operating and menu development. Additionally, the paper makes a discussion on the conceptual relations between menu and meal experience. Given the scarcity of research that incorporates evidences and concepts of previous studies in one single study, the conceptual structure of menu management presented in this paper allows a comprehensive understanding of menu and forms a theoretical basis for future research.

Keywords: Menu; Menu management; Restaurant firms

INTRODUCTION

Menu maintains its dominant position in restaurant firms, since it is the core of food and beverage operations. Specifically, menu is an instructor that clearly dictates (i) what will be produced, (ii) what type of equipment and ingredients are needed, and (iii) which qualifications employees should have. Menu also functions as a communicating and selling tool (Kincaid and Corsun, 2003). More specifically, it communicates not only the food and beverage offerings, but also the image of the firm. Furthermore, menu is a base on which the customers make their food choices and a well designed menu can direct customers' attention to the items the firm wants to sell more (Antun and Gustafson, 2005).

The critical qualifications attributed to menu describe its importance to restaurant operations and customer experiences. In fact, in contemporary understandings of menu, it has material and immaterial meanings. Materially, menu is a list or a card which documents the food and beverage options being offered by a restaurant. Immaterial meanings of menu highlight that it is a medium that affects customers' perceptions of restaurant experience (Wansink et al., 2005), and it is also an element of restaurant management that requires conducting critical managerial activities such as planning, pricing, designing and analyzing (Jones and Mifli, 2001). It also differs from the main product of a restaurant. Prior research (Johns and Kivela, 2001, Gustafsson, 2004 and Hansen et al., 2005) claim that the primary product of restaurants is meal experience and it is constituted by many components such as food and beverages, atmosphere, social factors and management. What customers receive from consumption in restaurants is their memories and overall perceptions of meal experience, and menu is part of that experience and has associations with customers' perceptions. Consequently, menu cannot be considered as the overall product of a restaurant but it is an important component complementing atmosphere, interactions with servers and other customers, and managerial systems which are jointly available in a restaurant where a good, memorable meal experience is the main product.

Since it is a key factor in attracting and satisfying customers and effecting performance of restaurant firms, menu has been gaining a considerable academic attention from researchers. The previous studies on menu constitute a substantial body of literature and they primarily examined the issues including menu planning, menu pricing, menu designing, and menu analysis. For instance, Kivela (2003) investigated menu planning through employing an experimental research design and proposed a menu item selection framework. Menu pricing was examined by several authors from a range of perspectives including the price perceptions of customers (Shoemaker, 1994 and Iglesias and Guillen, 2002), price sensitivity, pricing methods (Kelly et al., 1994, Raab et al., 2009 and Kelly et al., 2009), and psychological pricing (Carmin and Norkus, 1990, Naipaul and Parsa, 2001 and Parsa and Hu, 2004). Menu designing has also gained a particular attention from menu researchers who have particularly focused on positioning the menu item on the menu card (Bowen and Morris, 1995, Kincaid and Corsun, 2003 and Reynolds et al., 2005), labeling (Lockyer, 2006 and Wansink et al., 2001) and describing menu items (McCall and Lynn, 2008 and Hwang and Lorenzon, 2008). In the field of menu analysis, researchers have offered several models such as menu engineering (Kasavana and Smith, 1982), analysis of menu items' profitability (Hayes and Huffman, 1985), and more recently multi-dimensional analysis of menu items (Taylor and Brown, 2007). From a practical perspective, menu is directly related to many functions of food and beverage operations and also absorbs a wide range of interrelated managerial decisions in the areas of planning, pricing, designing, operating, and developing menus.

Despite the growing academic attention to restaurant menus, previous research has a certain drawback. Principally, there is an apparent lack of an attempt at incorporating the major thoughts and ideas of menu research into one single study. Given the above background, this conceptual study presents a review of relevant literature to frame the current discussion on menu with an attempt to identify critical managerial issues. Thus, the specific objectives of the current study are (i) to identify the critical managerial issues of menu, and (ii) to demonstrate the conceptual structure of menu management. As the menu is a key to influencing success of firms in foodservice industries, this study could assist restaurateurs in understanding the role of menu in their managerial decisions. From an academic perspective, this paper presents a holistic framework for menu researchers since it incorporates the conceptual and empirical findings of previous menu research into one single study.

Literature review

Theoretical underpinnings

The perspectives of organizational, marketing and strategic management theories provide a reliable theoretical groundwork to understand the important managerial aspects of menu. For instance, organization theory explicitly emphasizes the influence of external environment both on the decisions of firm managers and the survival of firms on the long run. More specifically, external environment is one of the central themes of organization theory and the relevant studies (Duncan, 1972, Jurkovich, 1974, Dess and Beard, 1984 and Ashill and Jobber, 1999) primarily propose that managers essentially consider the influential external factors that create uncertainty, diversity and volatility while making their decisions. Planning and operating menus in a restaurant context involve considering external factors such as customers, rivals, and vendors that have a great potential in creating uncertainty, diversity and volatility in the restaurants' immediate business environment. Complementing this view, marketing theory recognizes the importance of identifying the needs and expectations of customers, and developing and improving products and service perfectly fit to those needs and expectations. Additionally, pricing, promoting and distributing the products and services should also be consistent with the customers' needs and expectations on the one hand, and with the firm's objectives on the other hand. Thus, in the restaurant context, it is imperative that menu as the food and beverage combinations offered by a restaurant reflects the expectations and needs of customers. Moreover, managing menus involves planning, pricing, designing, distributing and promotional decisions which are also the main issues of marketing.

Strategic management is another promising theoretical perspective that helps us understand the importance of menu and its associated managerial activities with reference to competitiveness of restaurants. Indeed, creating and sustaining a competitive advantage is the main focus of strategic management discipline. More specifically, strategic management principally investigates the phenomenon of sustainable competitive advantage as a source of performance differences among business firms (Nerur et al., 2008). Following this argument, it is plausible to consider the role of menu in creating sustainable competitive advantage for restaurants. In particular, offering a unique meal experience and assuring customer satisfaction are the main sources of competitive advantage for a restaurant. As one of the important components of meal experience, menu deserves a special attention to understand how a restaurant firm can achieve its strategic goals, can create a sustainable competitive advantage, can satisfy its customers and can outperform the rivals.

Strategic management literature currently encompasses a wide range of strategic management schools posing different strategy definitions and processes (Sarvan et al., 2003). As a result, there are various understandings of the meaning of strategy (Okumus, 2002). For example, Mintzberg et al. (1998) define 10 schools of strategic thinking as design, planning, positioning, entrepreneurial, cognitive, learning, power, cultural, environmental and configuration schools. Among the 10 strategy schools, positioning school provides a clear typology of strategy alternatives. For instance, Porter (1980)proposes well-known three generic strategies: low cost leadership, differentiation and focus strategies, those are also available for restaurant firms. When considered in a restaurant context, each strategic alternative has associated implications for menu. For example, if the strategy is low cost leadership, then decisions on planning, pricing and designing menu reflect the strategy through selecting low cost menu items, pricing menu items at a lower price than rivals, and attractively presenting menu items with low prices on menu card, which probably result in customers' perceptions of a satisfactory meal experience with a reasonable quality. Conversely, when a differentiation strategy is chosen, a restaurant manager is expected to be attentive to (i) selecting or innovating unique menu items, (ii) setting higher prices with an expectation that customers are willing to pay premium for a differentiated product, and (iii) heavily concentrating on attractively presenting unique items on menu card to create a positive image of a unique meal experience. Thus, all managerial decisions with reference to planning, pricing and designing menus are the reflections of the chosen strategy such as low cost leadership or differentiation.

Menu management issues

Menu management is a relatively new concept which is not widely recognized by researchers or practitioners. Indeed, in the relevant literature very few studies have used the concept without providing a certain definition of it. Thus, the major aim of literature review in this section is to trace the variables investigated by menu researchers. The review purposely restricted to the journal articles which were subjected to a rigorous academic judgment process before publishing.

Table 1. Examples of articles investigating menu issues.

Article	Main focus	Corresponding menu issue
Pavesic (1983)	Menu engineering model including weighted contribution margin and food cost percentage	Menu analyzing
Hayes and Huffman (1985)	Menu item profitability analysis and goal value analysis	Menu analyzing
Bayou and Bennett (1992)	Menu item profitability analysis based on analysis of market segments (four levels of profitability analysis)	Menu analyzing
Kelly et al. (1994)	The relationship between price elasticity and pricing	Menu pricing
Shoemaker (1994)	The effect of a decoy item on the price value perception of a menu	Menu pricing
Bowen and Morris (1995)	The relationship between menu item placement on the menu card and sales of menu items	Menu designing
LeBruto et al. (1995)	Menu engineering model including labor costs as a menu analysis method	Menu analyzing
Morrison (1997)	Development of menu engineering models and a study on upscale restaurants menu planning practices	Menu analyzing Menu planning
Jones and Mifli (2001)	Menu planning, pricing, designing, analyzing and improving practices of seven hospitality firms	Menu planning Menu pricing Menu designing Menu analyzing and improving
Naipaul and Parsa (2001)	The relationships between psychological pricing comprising odd and even number price endings, and value and quality perceptions of customers	Menu pricing
Wansink et al. (2001)	The relationship between descriptive menu item labels and customer perceptions	Menu designing
Iglesias and Guillen (2002)	Price perceptions of customers comprising monetary and non-monetary sacrifices	Menu pricing
Kincaid and Corsun (2003)	The relationship between the layout of menu item on menu card and item sales	Menu designing
Kivela (2003)	Menu planning approaches and menu item selection variables	Menu planning
DiPietro et al. (2004)	Consumer attitudes towards healthy menu items and investigation of sales trends in healthy menu items	Menu planning
Antun and Gustafson (2005)	Menu design characteristics and pricing strategies used by restaurants and clubs	Menu pricing Menu designing
Kwong (2005)	Asian restaurants' practices of menu engineering and menu design	Menu analyzing Menu designing
Cohen et al. (2006)	The menu item dependency and the effect of loss-leader pricing on menu items performances, and proposal of a multidimensional menu analysis technique	Menu analyzing
Lockyer (2006)	The relationships between the wording of menu items and customers' menu item choice	Menu designing
Taylor and Brown (2007)	A review of menu analysis literature including improvements to menu item performance	Menu analyzing and improvements
Ottenbacher and Harrington (2007)	The menu item innovation process in fine dining restaurants	Menu planning
Bernstein et al. (2008)	The relationship between menu variability and consumers' restaurant preferences	Menu planning
Hwang and Lorenzon (2008)	Relationships among nutrition labeling of menu items, consumer perceptions and menu pricing	Menu pricing Menu designing
McCall and Lynn (2008)	The relationship between the complexity of menu item descriptions and potential customers' evaluations	Menu designing
Raab et al. (2009)	Pricing approaches adopted by hospitality firms and price sensitivity measurement as a pricing method	Menu pricing
Taylor et al. (2009)	Proposal of data envelopment analysis as menu analysis technique	Menu analyzing
Ottenbacher and Harrington (2007)	The menu item innovation process in quick service restaurants	Menu planning

The previous research includes empirical and conceptual studies examining various managerial aspects of menu. Table 1 demonstrates the examples of those studies and their variables. As seen in Table 1, there are different sets of variables corresponding to one or several of the menu management issues including menu planning, menu pricing, menu designing, and menu development. Indeed, among the articles referred in Table 1, the empirical study of Jones and Mifli (2001), which also forms the conceptual base for the current study, jointly investigates planning, designing, developing, pricing and analyzing issues in seven hospitality firms. Jones and Mifli (2001: 61–62) relate menu planning to identifying the product range of restaurant firm, and menu design to the creation of menu card. They define menu development as subsequent alteration of the planned menu. Finally, menu analysis is associated with a range of techniques and procedures employed in order to analyze the menu performance. The following section elaborates on menu management issues by drawing evidence from the menu literature.

Menu planning

Recent studies on menu planning examine variables such as menu item selection process and criteria, menu changes and variety, and menu item innovation. Those studies primarily assume that menu planning is a critical managerial activity to the success of restaurant firms. In this regard, Kivela (2003) describes menu planning as a menu item selection process. After heavily criticizing the traditional approach to menu planning which mainly relies on executive chefs' gastronomic knowledge, he proposes a Menu Planning Qualitative Variables (MPQV) approach which considers gastronomic, financial, marketing and menu mechanics variables. Morrison (1997) and Kwong (2005) also share the idea that menu planning is a process of selecting menu items and their research reports empirical evidences that the criteria of menu item selection are based on the cost of raw materials, cost of labor, profitability of menu item, the skill level of staff needed, availability of ingredients, space and equipment requirements, and customer demand. A valid alternative to selecting menu items from an existing items' pool based on some pre-determined criteria is to make menu item innovations. Recently, some researchers (Ottenbacher and Harrington, 2007 and Ottenbacher and Harrington, 2009) have focused on innovation processes in restaurant kitchens. The studies of those researchers empirically demonstrate how chefs create new menu items relying on innovation processes that are generally encompassing the stages such as idea generation, screening, concept development and tests, commercialization, implementation and evaluation.

Glanz et al. (2007) emphasize that changing menu's content is also an important component of menu planning and their study reveals that the major criteria of menu changes are whether the new menu item will attract new customers or maintain the existing customer base, and how it will affect sales and profit. Furthermore, menu changes may lead to menu variety, and according to findings presented by Bernstein et al. (2008), consumers value variety in their menu choices. After analyzing the recent trends in food preferences of consumers, DiPietro et al. (2004) conclude that healthy food items are becoming an increasingly important part of restaurant menus even in quick service segment. With respect to the growing trend towards healthy food choices, menu planning becomes more complicated due to associated activities for searching, selecting or innovating the healthy menu options.

Menu planning researchers have contributed to establishing the conceptual structure of menu planning by introducing and exploring menu item selection, menu item innovation, menu variety, menu change and menu content variables. It seems that research on menu planning largely holds management theories' perspective and highlights the effects of external (customer, rivals, vendors) and internal (cost structure, employees, equipment) factors on menu planning decisions of managers. Considering the findings and suggestions of the previous studies, it is possible to make an operational definition of menu planning. In this regard, menu planning refers to menu item selection and/or menu item innovation with a consideration of menu variety and menu change in order to generate a menu content of an optimum composition of food and beverage offerings that are most likely to gratify both the customers' and restaurant firm's expectations.

Menu pricing

Menu pricing is a quite popular topic among menu researchers and it seems that a wide range of variables such as price perceptions of customers, price sensitivity, price elasticity, pricing methods and specifically the psychological pricing have attracted their attention. For instance, Iglesias and Guillen (2002) specifically underline the importance of customers' perceptions of menu item prices and suggest that total price perception of customers consists of perceived monetary price and perceived non-monetary price. Additionally, they indicate that customers also consider reference price and compare it to the actual price of menu item when making purchase decisions. Specifically, to some extent the assessment of the price of a menu item is based on the prices of alternatives. In a restaurant setting, for example, the prices of decoy items on the menu card may act as reference prices (Shoemaker, 1994). Depending on the price sensitivity or price elasticity, customer demand to food and beverage offerings of a particular restaurant may change in response to an increase or decrease in the prices (Kelly et al., 1994 and Raab et al., 2009). Moreover, presentation of price endings can affect customers' value or quality perceptions of associated menu items when they are exposed to psychological pricing methods (Naipaul and Parsa, 2001).

The researchers in the field of menu pricing install a marketing theory perspective into menu management through emphasizing the perceptions and expectations of customers. Following the conceptual background provided by previous research, menu pricing can be described as the formulation of menu item prices in response to both sales, cost and profit objectives of the firm, and perceptions and expectations of customers. Additionally, as suggested by the relevant literature, setting the menu item prices is a complicated task since it requires becoming attentive to complex effects of the interrelated factors such as customers' perceptions, price elasticity and psychological pricing.

Menu designing

Researchers suggest that menu designing comprises managerial decisions regarding design of menu card characteristics, position, label, and description of menu items on the menu display (card or board). More specifically, the previous research on menu design has associated the menu item's position, label and description with item sales and customers' perceptions.

Although there are theoretical explanations asserting that the menu card contains some sweet spots and a particular menu item placed on a sweet spot generates higher sales; the empirical evidences reported by Bowen and Morris (1995), Kincaid and Corsun (2003), Reynolds et al. (2005) do not support that assumption. However, recently Choi et al. (2010) and Yang (2012) report evidences that restaurant menus may consist of sweet spots where the first gazes of customers focus on. Some other researchers present empirical findings that verify the association between an item's sales and its label and description. For instance, the findings of Wansink et al. (2001) and Lockyer (2006)evidence that labeling or wording of menu items' names significantly influences the customers' menu item choice and perceptions. More specifically, according to Wansink et al. (2005), evocative labels on a menu card attract customers' attention to specific items and positively affect their taste and value perceptions of those items. Additionally, the empirical study of McCall and Lynn (2008) demonstrate the importance of item descriptions on a menu card through the finding that the complexity of menu item descriptions positively affects the customers' evaluations. Shoemaker et al. (2005)present a similar finding and conclude that details provided in items' descriptions have a positive impact on customers' value perceptions. Alternatively, Hwang and Lorenzon (2008) advocate that nutritional information provided on the menu card can enhance the positive attitudes of customers by informing them about healthy options in the menu and further make them change their orders to healthy items. Menu design also relates to a variety of factors including background and text colors, texture, photos, fonts, the use of boxes, and menu size as characteristics of menu card (Reynolds et al., 2005) which also have considerable effects on customers' item choice as empirically supported byGuéguen et al. (2012) who found that pictures of sea on the menu card significantly increase the consumption of fish dishes.

The studies focusing on menu design seem to make considerable contribution to our understanding of menu management through clarifying the relationships between item sales and menu design variables. In this way, they introduce a merchandising approach to menu management and hold a marketing perspective which assists managers in attracting customer attention to particular menu items. Given the above background, menu design refers to designing menu card with respect to menu card characteristics, menu item layout, menu item descriptions and labels in order to create an attractive menu that not only provides information but also directs customers' attention to the items that the firm wants to sell more.

Menu operating

Menu operating is essentially included in menu management issues despite the fact that it very rarely appears in Table 1. In fact, it is not possible to state that menu researchers have paid particular attention to operations of menu except Jones and Mifli (2001). However, the studies on subjects such as sales forecasting, budgeting, cost control, service failures, and food safety provides insightful contribution to our understanding of menu operating.

In particular, menu operating involves management of menu items' production and service processes. Thus, it is expected that restaurant managers frequently consider three key factors as quality, hygiene and cost while making decisions and taking actions of menu operating. Accordingly, standardizing production, minimizing service failures, assuring food safety, forecasting demand to menu items, making budgets and controlling costs become critical activities of operational management of menus. These are also the major considerations of quality management approaches such as ISO: 9001 and ISO 22000 which are also adaptable to restaurant operations. When considered in the restaurant context, one of the main components of such a quality management system will be assurance of food quality and safety relying on standard production, minimum service failures, optimum demand and suitable cost structures; those are briefly explained below with reference to relevant literature.

In the production process of menu items, development and implementation of some standards such as standard yields, standard purchasing specifications, standard recipes and standard portions are useful tools for sustaining a quality level (Davis and Stone, 1998). As found by Jones and Mifli (2001), especially in the chain restaurants portion control, standard recipes, quality control, and standard specifications are integral components of menu operations.

Serving a meal in a restaurant (especially in a table service restaurant) is composed of several sequential service stages such as greeting and seating, order taking and delivery, consumption, and payment and exiting (Namkung et al., 2011). Inevitably, service failures may occur at any stage of service, and these failures may lead to negative customer perceptions. For example, customers may complain following a failure (Namkung et al., 2011). Service failures may also negatively affect the customers' satisfaction levels and their likelihood of returning (Namkung and Jang, 2010). Consequently, menu operating decisions of managers need to consider serving menu items properly without any delays or defects in order to minimize service failures at all service stages.

Restaurant customers are becoming increasingly concerned about possible risks stemming from unsafe food practices (Fatimah et al., 2011) and food safety is a critical restaurant selection attribute (Worsfold, 2006). Thus, in menu operating stage restaurant managers are challenged by an urgency to produce and serve safe food. In order to ensure food safety, restaurant managers need to recognize the potential biological, physical and chemical hazards and take preventive actions to minimize the occurrence of these hazards (Fatimah et al., 2011). However, restaurant industry frequently faces food safety problems that mainly occur due to unsafe food sources, inadequate cooking, improper food handling, improper storage temperatures, contaminated equipment and poor personal hygiene (Sanlier et al., 2010). As a solution to the current food safety problems, it is widely advised that restaurant firms should implement a food safety management system such as HACCP (Hazard Analysis and Critical Control Points) which consists of principles and practices that encompass all aspects of processing, preparing, handling and serving food in a completely safe way (Eves and Dervisi, 2005).

Sales forecasts have a key role in operational management of menu due to the perishable nature of food products. More specifically, an accurate prediction of demand to menu items ensures the smooth functioning of production and service without falling behind the demand, which may cause customer dissatisfaction, or exceeding the demand, which increases the costs (Ryu et al., 2003). Previous studies on sales forecasting have documented the available forecasting methods such as simple moving average method, simple exponential smoothing method, time-series models, regression, and Delphi etc. (Green and Weaver, 2008). Some researchers suggest the use of those quantitative techniques rather than merely intuitive assessments of managers since naïve intuitive approaches produce less accurate forecasts than do quantitative techniques (Miller et al., 1991). Besides, Ryu et al. (2003) also highlight the importance of using simple and timesaving forecasting methods.

In menu operating, budgets and cost control systems assist restaurant managers in assuring that there is no serious deviation from profit and cost objectives of restaurant firm in the entire process of procurement, storage, production and service. Underlying this notion, research by Jones (2008) evidences that hospitality firms produce budgets for cost control and performance evaluation purposes. Loss of profit due to the poor controls is a common concern in hospitality settings (Borchgrevink and Ancill, 2003). As a result, restaurant managers consider and implement a cost control system which encompasses purchasing, storage, issuing, production and service stages. The major aim of such a system is to minimize potential loss because of shrinkage, spoilage, spillage, waste and theft. However, as stated by Vaughn et al. (2010), the cost control system of many restaurant firms is based solely on the control of food costs. As a result, the restaurant has no knowledge of ultimate cost of menu items. Alternatively, Vaughn et al. (2010) suggest the use of activity based costing method in order to accurately determine and control the cost of menu items.

Given the above explanations, menu operating refers to smooth operation of menu with assistance of sales forecasts, standards, food safety practices, budgets and cost control systems in order to minimize deviations from quality, hygiene and cost objectives of restaurant firms.

Menu analysis

Researchers have particularly dealt with developing some quantitative models to analyze the performance of menu items. Additionally, some of them also proposed improvement strategies to eliminate the performance deficiencies of menu items. In the 1980s, early systematic menu analysis approaches based on matrixes were introduced by Miller (1980), Kasavana and Smith (1982) and Pavesic (1983). These matrix based models were criticized by subsequent studies and several weaknesses were noted (Taylor and Brown, 2007). Among them is the notion that indicator of menu items' profitability is only the contribution margin which is computed through subtracting food cost from selling price. Consequently, this approach does not include other variable and fix costs such as labor and energy. Paying special attention to this deficiency of matrix models, LeBruto et al. (1995) attempt to include labor costs into the menu engineering model. Moreover, Hayes and Huffman (1985) and Bayou and Bennett (1992), known as the developers of menu item profitability analysis, seek to identify profit of menu item by subtracting total cost from sales of the item. Recently, some multi-dimensional analysis approaches were developed (Taylor and Brown, 2007). For instance, Cohen et al. (2006)propose a multi-dimensional menu analysis model employing five performance dimensions including food cost, popularity, sales price, contribution margin, and weighted contribution margin. More recently, Taylor et al. (2009) suggest that data envelopment analysis is a useful multidimensional technique to analyze menu item performance.

Taylor and Brown (2007), Jones and Mifli (2001) and Kwong (2005) provide information about possible improvement strategies for menu item performance enhancement. Indeed, subsequent to the analysis of menu is the implementation of suitable improvement strategies for performance deficiencies. According to the authors, promotion and reposition of menu item; modification of presentation, recipe, and price are strategies for increases in the sales volume. Recosting, repricing, and recipe modification are optional solutions to cost problems. High performer menu items may be left as they are and low performers may be eliminated.

Studies on menu analysis reveal that evaluation of menu performance is mainly based on the financial performance indicators such as cost, popularity, profit and price. Additionally, improvements are made in order to increase sales or decrease costs. Therefore, it is obvious that menu analysis is made with respect to restaurant firms' financial goals. Relying on the arguments in the relevant literature, menu development may be proposed as a two dimensional concept including menu analysis and improvements instead of menu analysis concept alone. In this sense, menu development refers to analyzing menu item performance and then taking corrective actions to improve the performance imperfections of menu items.

DISCUSSION

It becomes clear that prior menu research has offered major menu management issues such as menu planning, menu pricing, menu designing, menu operating, and menu development. Critically, these menu management issues are not independent but interrelated. The success of any decision made in relation to a particular menu management issue largely depends on the performance rendered in relation to other issues. Furthermore, menu management is not a static managerial activity, but it is an ongoing process. This process has at least two main objectives including (i) satisfying the customers, and (ii) achieving the goals of restaurant firm. As stated by Kincaid and Corsun (2003), the menu contributes to restaurant firm's performance through generating profit for the firm and providing value and satisfaction to the customer. In this regard, menu management is a comprehensive concept that refers to an ongoing process comprising planning, pricing, designing, operating and developing the menu of a restaurant firm. Table 2 summarizes the conceptual structure of menu management along with the relevant variables extracted from literature.

Table 2. Conceptual structure of menu management.

Management issue	Short description	Outcome	Relevant variables
Menu planning	Process of generating an optimal menu content	Menu content	Menu item selection criteria Menu changes and variety Menu item innovation
Menu pricing	Formulation of menu item prices	Menu item prices	Pricing methods Price elasticity or sensitivity Price perceptions of customers Psychological pricing
Menu designing	Designing the menu display (menu card or board)	Menu display (menu card or board)	Menu card characteristics Menu item layout Menu item description Menu item label
Menu operating	Developing and implementing standards and procedures regarding quality, hygiene and cost	Operations	Budgeting Sales forecasting Assuring food safety Controlling costs Standard recipes Standard costs Standard yields Standard portions
Menu development	Analyzing performance of menu items and improving the performance imperfections	Menu items' performances and improvements	Analysis methods Improvement strategies

As seen in Table 2, planning, pricing, designing, operating, analyzing and improving menu include managerial decisions. Those decisions may have profound effects on customers' positive perceptions of the meal experience and their satisfaction with restaurant which further lead to a better performance of the restaurant firm. As meal experience is the primary product of restaurant firms (Johns and Kivela, 2001), then an assessment of associations between menu management issues and components of meal experience may contribute to our understanding of how menu might enhance the meal experience perceptions of customers.

A synthesis of existing literature (Gustafsson, 2004 and Hansen et al., 2005) reveals that components of meal experience are restaurant interior and atmosphere, social meeting, food and beverages, company of other people, and management control system. Since food is one of the most influential restaurant selection attributes (Kivela, 1997) and a determinant of customer satisfaction and retention (Kivela et al., 2000), food and beverages constitute the core of meal experience (Gustafsson, 2004). As menu management determines what will be the combination of food and beverages and their prices, how they will be presented on the menu card, and how they will be produced and served, it is possible to assume that menu has primary implications for the core component of meal experience. For instance, a menu planning activity attentive to customer expectations assures that customers can receive what they expect in terms of menu content, menu variety and changes. Similarly, an efficient use of menu pricing approaches can affect the way customers perceive the meal experience. Certain practices of presenting price-endings and considering customers' price perceptions while establishing prices play a key role on how customers evaluate the menu items and restaurant itself. For instance, in quick service settings price endings with "9" may generate a sense of value while "0" in price-endings may create a sense of quality in full-service settings (Naipaul and Parsa, 2001). Moreover, differentiating the menu items and accordingly their prices from potential competitors' offerings through innovations in dishes (Ottenbacher and Harrington, 2009) may assist in forming the customers' perception of a unique meal experience without any reference to prices of rivals' products (Iglesias and Guillen, 2002).

Existing menu design research suggests that customers' purchasing behavior and value and quality assessments can also be influenced by menu item labeling and description (Yang et al., 2009). For instance, menu item labels which awake positive emotions of customers, and descriptions including nutritional information may enhance the meal experience perceptions of customers. Specifically, labels and descriptions of menu items may lead customer to have positive perceptions of taste, value, health and quality aspects of food (Shoemaker et al., 2005; Wansink et al., 2001 and Wansink et al., 2005). Additionally, a restaurant manager who wishes to ensure that customers have a unique meal experience may consider that the unique food and beverage offerings of the establishment are located on the menu card where they are most probably seen and frequently ordered (Choi et al., 2010). Moreover, physical characteristics of menu card such as background, colors, texture, photos, fonts, size etc. may be designed in combination with the other atmospherics of the restaurant (décor, lighting, layout, table settings, appearance of service staff etc.) in order to generate an overall appealing image as the restaurant atmospherics have a considerable effect on customers' perception of meal experience, and their satisfaction and retention (Ryu and Han, 2011).

Operational decisions and actions related to menu such as standardizing and offering a consistent quality, minimizing service failures, accurately predicting demand to menu items, and assuring food safety are essential components of providing a prosperous meal experience. For instance, customers may be dissatisfied when their orders are delayed or cannot be served due to shortage of ingredients because of poor sales forecasting. Moreover, customers who are attentive to food safety issues may assess all visible cues such as cleanliness of utensils, serving temperature and personal hygiene of service staff in order to be certain the safety level of food and these assessments may influence their satisfaction with meal experience (Fatimah et al., 2011).

The management control system component of meal experience also deserves attention from menu management perspective. Sales, cost and profit are subjected to consideration in menu management. For instance, selecting menu items with reference to sales; cost and profit objectives of firm; increasing item sales with pricing and designing tactics; controlling the costs with assistance of sales forecasting, and budgeting and cost control systems are all related to management control system component of meal experience. Menu analysis also assists in management control system. As a periodical performance monitoring mechanism, menu analysis is undertaken in order to evaluate whether the menu management process is successful. Menu analysis requires recording and collecting menu items' performance data which are specific to the establishment and those data can be integrated into the restaurant's overall performance assessments. Menu analysis also contributes to enhancing the meal experience when it includes improvements in terms of making adjustments in menu planning, pricing, designing and operating in response to customers' expectations.

CONCLUSION

Relying on the previous studies on menu, this study shows that the menu literature is constituted by five menu management issues comprising menu planning, menu pricing, menu designing, menu operating and menu development. Menu management concept, then, is defined as an ongoing process including menu management issues.

Managerial implications

A review of menu research shows us that restaurant managers have to play diverse and inter-related roles in relation to menu management. In this sense, a restaurant manager should become a planner, a price setter, a designer, an operator and an analyst. On the one hand, these roles are diverse because each of them requires different tasks listed inTable 2 in this study. On the other hand, they are interrelated since each task should be accomplished in close relation to others. Thus, the challenging aspect of menu management is that managers essentially integrate the diversified and complex menu management issues and understand the interdependencies among them instead of treating each issue individually. In this way, managers may have an ability to use menu in enhancing the meal experience perceptions of customers and attaining the cost and profit objectives of restaurant firms.

As discussed in the current paper, menu may have a strong influence on customers' meal experience perceptions. The restaurants which aim to have a higher level of performance than their rivals essentially concentrate on creating a unique meal experience. Meal experience is a multi-component construct and restaurant managers need to find unique ways of combining the meal experience components and menu variables in order to provide a unique experience for their customers. Relying on the literature review presented above, it seems that one possible way to accomplish this is to treat menu management as a process consists of interconnected stages and following paragraphs include some major suggestions to managers explaining how to operate such a process.

Since, menu is a determinant of purchasing, receiving, storage, production and service requirements, a starting point for managers is to identify the current state of internal and external factors related to menu management. For instance, the equipment and staff portfolio of the firm in terms of qualitative and quantitative features should be overviewed because of the strong influence of menu on the equipment and staff requirements. The processes, cost structures, profit, sales and prices of rivals should be examined in order to assess and differentiate the firm's own offerings. Importantly, the needs and expectations of the existing and potential customers need to be captured and analyzed. Other environmental factors such as economical, social, political and legal issues and the possible changes in them should also be analyzed. In this regard, at the initial stage of menu management process there should be a situational analysis that comprises collecting and analyzing data related to external and internal variables such as customers, rivals, environmental factors, and the firm's capabilities. Consequently, the situational analysis stage forms a base which contains useful data to effectively perform the subsequent menu planning, pricing and designing stages of menu management process.

As stated by Reich (2002), a restaurant firm may have long term and short term objectives. The long term objectives comprise social responsibility, profitability, productivity, employee development, employee relations, competitive position and technology development. Short term objectives include increasing profitability, decreasing costs, increasing customer satisfaction, and increasing market share. Accordingly, it is reasonable to state that the menu management process should produce outcomes which are consistent with the long term and short term objectives of the restaurant firm. A review of the firm's long term and short term objectives provide insights to understand the expected outcomes of the menu management process. Thus, in the second stage following the situational analysis, managers should clearly identify the specific objectives of menu management process. The objectives specific to menu management are most likely to be the short term objectives such as profitability, cost levels, and customer satisfaction.

Menu planning, pricing and designing are interrelated issues. In order to accomplish the task of pricing, managers need to know the menu items, their contents and accordingly their costs. Similarly, menu designing requires knowing the names, content and prices of menu items. Furthermore, managers should base their menu planning, pricing, and designing decisions on the outcomes of situational analysis and review of objectives.

Subsequent to planning, pricing and designing, the planned menu is put into production. Therefore, operating the planned menu requires adopting standard procedures in order to achieve the expected performance determined in the planning stage. The menu operation should also consist of sales forecasting, food safety, budgeting, and cost control. Then, the aim of operation stage is to guarantee that actual performance of menu is consistent with the expected performance.

Finally, menu development corresponds to strategic evaluation since it comprises analysis and improvements. Menu development allows managers to monitor and analyze the menu performance periodically and to make improvement decisions relying on the performance analysis. The improvement decisions may aid establishing a better alignment between the external environment and internal functioning of the firm which means a better strategic fit. In this way, menu development produces feedback to menu planning, pricing and designing. The feedback from menu development stage is also an indicator of deficiencies in the menu operation in terms of deviations from standard procedures. Moreover, the information provided by menu development may indicate to make some adjustments to the objectives of firm.

Limitations and suggestions for future research

The conceptual structure of menu management presented in this study provides a tool for hospitality researchers to investigate the menu issues. Researchers can use this conceptual structure as an organizing tool in order to have an overall view of menu management concept and process. For instance, the relationships among menu management issues need to be examined through employing empirical research design. Theoretically, each step of menu management process is an essential preceding step for next steps. Additionally, the efficacy of each menu management issue positively relates to performance of menu management and overall performance of restaurant firm. However, these conceptual relationships need to be empirically investigated. Although this study has made an effort to conceptually understand the associations between menu management issues and meal experience, there is another need for empirical research investigating the use of menu in enhancing the meal experience. Consequently, some guiding propositions may be derived from the menu management process. First, menu management issues are interrelated. Second, the effectiveness of menu management issues is positively related to the menu performance. Third, the effective management of menu is positively related to the overall performance of restaurant firms. Finally, menu management issues influence the meal experience perceptions of customers.

This is a conceptual study attempting to identify conceptual structure of menu management. Therefore, the major limitation of the study is that it cannot present empirical evidences which are obtained through a specific research design. However, an extensive literature review was conducted in order to reflect the theoretical and empirical insights from the relevant previous studies. Another limitation of the current paper is that the menu management process defined in this study is a general one and ignores the specific changes that may occur in different segments of restaurant industry. That is why an empirical research study which focuses on a comparative analysis of menu management issues in different segments of restaurant industry such as full service versus quick service is needed and such an empirical study will probably produce very useful findings that enhance our understanding of restaurant menus. Additionally, an interesting empirical investigation might be the exploration of differences in menu management issues depending on the different times of consumption such as management of lunch or dinner menus, business-day menus or seasonal menus. In fact, the time when the menu is consumed by customers probably requires different strategies, but because of the space constraints it was not possible to explain those strategies in the current paper. In conclusion, this study fills a gap in the menu literature through offering an integrative conceptual structure of menu management.

REFERENCES

1. Antun, J.M., Gustafson, C.M., 2005. Menu analysis: design, merchandising and pricing strategies used by successful restaurants and private clubs. J. Nutr. Recipe Menu Dev. 3 (3), 81–101.
2. Ashill, N.J., Jobber, D., 1999. The impact of environmental uncertainty, decision-maker characteristics and work environment characteristics on the perceived usefulness of marketing information systems (mkis): a conceptual framework. J. Mark. Manag. 15, 519–540.
3. Bayou, M.E., Bennett, L.B., 1992. Profitability analysis for table service restaurants. Cornell Hotel Restaur. Adm. Q. 33 (2), 49–55.
4. Bernstein, D., Ottenfeld, M., Witte, C.L., 2008. A study of consumer attitudes regarding variability of menu offerings in the context of an upscale seafood restaurant. J. Foodserv. Bus. Res. 11 (4), 398–411.
5. Borchgrevink, C.P., Ancill, D.M., 2003. Beverage control systems and the guests' perspective: standardization and value. J. Hosp. Leis. Mark. 10 (1/2), 151–159.
6. Bowen, J.T., Morris, A.J., 1995. Menu design: can menus sell? Int. J. Contemp. Hosp. Manag. 7 (4), 4–9.
7. Carmin, J., Norkus, G., 1990. Pricing strategies for menus: magic or myth? Cornell Hotel Restaur. Adm. Q. 31 (3), 7–44.
8. Choi, J.G., Lee, B., Mok, J., 2010. An experiment on psychological gaze motion: a re-examination of item selection behavior of restaurant customers. J. Glob. Bus. Technol. 6 (1), 68–79.

9. Cohen, E., Ghiselli, R., Schwartz, Z., 2006. The effect of loss leader pricing on restaurant menus' product portfolio analysis. J. Foodserv. Bus. Res. 9 (1), 21–38.

10. Davis, J.D., Stone, S., 1998. Food and Beverage Management. Butterworth Heinemann, Oxford.

11. Dess, G., Beard, D., 1984. Dimensions of organizational task environments. Adm. Sci. Q. 29, 52–73.

12. DiPietro, R.B., Roseman, M., Ashley, R., 2004. A study of consumers' response to quick service restaurants' healthy menu items: attitudes versus behaviors. J. Foodserv. Bus. Res. 7 (4), 59–77.

13. Duncan, R.B., 1972. Characteristics of organizational environments and perceived environmental uncertainty. Adm. Sci. Q. 17, 313–327.

14. Eves, A., Dervisi, P., 2005. Experiences of the implementation and operation of hazard analysis and critical control points in the foodservice sector. Int. J. Hosp. Manag. 24, 3–19.

15. Fatimah, U.Z.A.U., Boo, H.C., Sambasivan, M., Salleh, R., 2011. Foodservice hygiene factors – the consumer perspective. Int. J. Hosp. Manag. 30, 38–45.

16. Glanz, K., Resnicow, K., Seymour, J., Hoy, K., Stewart, H., Lyons, M., Goldberg, J., 2007. How major restaurant chains plan their menu: the role of profit, demand, and health. Am. J. Prev. Med. 32 (5), 383–388.

17. Green, Y.N.J., Weaver, P.A., 2008. Approaches, techniques, and information technology systems in the restaurants and foodservice industry: a qualitative study in sales forecasting. Int. J. Hosp. Tourism Adm. 9 (2), 164–191.

18. Guéguen, N., Jacob, C., Ardiccioni, R., 2012. Effect of watermarks as visual cues for guiding consumer choice: an experiment with restaurant menus. Int. J. Hosp. Manag. 31 (2), 617–619.

19. Gustafsson, I., 2004. Culinary arts and meal science – a new scientific research discipline. Food Serv. Technol. 4 (1), 9–20.

20. Hansen, K.V., Jensen, Q., Gustafsson, I., 2005. The meal experience of a la carte restaurant customers. Scand. J. Hosp. Tourism 5 (2), 135–151.

21. Hayes, D.K., Huffman, L., 1985. Menu analysis: a better way. Cornell Hotel Restaur. Adm. Q. 25 (4), 64–70.

22. Hwang, J., Lorenzon, C.L., 2008. Effective nutrition labeling of restaurant menu and pricing of healthy menu. J. Foodserv. 19, 270–276.

23. Iglesias, M.P., Guillen, M.J.Y., 2002. The components of total perceived price: an empirical analysis in restaurant services. J. Foodserv. Bus. Res. 5 (1), 1–22.

24. Johns, N., Kivela, J., 2001. Perceptions of the first time restaurant customer. Food Serv. Technol. 1, 5–11.

25. Jones, P., Mifli, M., 2001. Menu development and analysis in UK restaurant chains. Tourism Hosp. Res. 3 (1), 61–71.

26. Jones, T.A., 2008. Improving hotel budgetary practice – a positive theory model. Int. J. Hosp. Manag. 27, 529–540.

27. Jurkovich, R., 1974. A core typology of organizational environments. Adm. Sci. Q. 19, 380–394.

28. Kasavana, M.L., Smith, D.J., 1982. Menu Engineering. Hospitality Publications Inc., Lansing, MI.

29. Kelly, T.J., Kiefer, N.M., Burdett, K., 1994. A demand-based approach to menu pricing. Cornell Hotel Restaur. Adm. Q. 50 (3), 383–387.

30. Kincaid, C.S., Corsun, D.L., 2003. Are consultants blowing smoke? An empirical test of the impact of menu layout on item sales. Int. J. Contemp. Hosp. Manag. 15 (4), 226–231.

31. Kivela, J.J., 1997. Restaurant marketing: selection and segmentation in Hong Kong. Int. J. Contemp. Hosp. Manag. 9 (3), 116–123.

32. Kivela, J.J., 2003. Results of a qualitative approach to menu planning using control and experimental groups. J. Foodserv. Bus. Res. 6 (4), 43–65.

33. Kivela, J.J., Inbakaran, R., Reece, J., 2000. Consumer research in the restaurant environment. Part 3: Analysis, findings and conclusions. Int. J. Contemp. Hosp. Manag. 12 (1), 13–31.

34. Kwong, L.Y.L., 2005. The application of menu engineering and design in Asian restaurants. Int. J. Hosp. Manag. 24, 91–106.

35. LeBruto, S., Ashley, R., Quain, W., 1995. Menu engineering: a model including labor. FUI Hosp. Rev. 13 (1), 41–50.

36. Lockyer, T., 2006. Would a restaurant menu item by any other name taste as sweet? FUI Hosp. Rev. 24 (1), 21–31.

37. McCall, M., Lynn, A., 2008. The effects of restaurant menu item description on perceptions of quality, price, and purchase intention. J. Foodserv. Bus. Res. 11 (4), 439–445.

38. Miller, J.E., 1980. Menu Pricing and Strategy. CBI Publishing, Boston.

39. Miller, J.L., McCahon, C.S., Bloss, B.K., 1991. Food production forecasting with simple time series models. Hosp. Res. J. 14 (3), 9–21.

40. Mintzberg, H., Ahlstrand, B., Lampel, J., 1998. Strategy Safari: A Guided Tour Through the Wilds of Strategic Management. The Free Press, New York.

41. Morrison, P., 1997. Menu engineering in upscale restaurants. Br. Food J. 99 (10), 388–395.

42. Naipaul, S., Parsa, H.G., 2001. Menu price endings that communicate value and quality. Cornell Hotel Restaur. Adm. Q. 42 (1), 26–37.

43. Namkung, Y., Jang, S., 2010. Service failures in restaurants: which stage of service failure is the most critical? Cornell Hosp. Q. 51 (3), 323–343.

44. Namkung, Y., Jang, S., Cho, S.K., 2011. Customer complaints in restaurants: do they differ by service stages and loyalty levels? Int. J. Hosp. Manag. 30 (3), 495–502.

45. Nerur, S.P., Rasheed, A.A., Natarajan, V., 2008. The intellectual structure of strategic management filed: an author co-citation analysis. Strat. Manag. J. 29, 319–336.

46. Okumus, F., 2002. Can hospitality researchers contribute to strategic management literature? Int. J. Hosp. Manag. 21, 105–110.

47. Ottenbacher, M., Harrington, R.J., 2007. The innovation development process of michelin-starred chefs. Int. J. Contemp. Hosp. Manag. 19 (6), 444–460.

48. Ottenbacher, M., Harrington, R.J., 2009. The product innovation process of quick service restaurant chains. Int. J. Contemp. Hosp. Manag. 21 (5), 523–541.

49. Parsa, H.G., Hu, S., 2004. Price-ending practices and cultural differences in the food service industry: a study of Taiwanese Restaurants. Food Serv. Technol. 4, 21–30.

50. Pavesic, D.V., 1983. Cost-margin analysis: a third approach to menu pricing and design. Int. J. Hosp. Manag. 2 (3), 127–134.

51. Porter, M.E., 1980. Competitive Strategy. Free Press, New York.

52. Raab, C., Mayer, K., Kim, Y.S., Shoemaker, S., 2009. Price-sensitivity measurement: a tool for restaurant menu pricing. J. Tourism Hosp. Res. 33, 93–105.

53. Reich, A.Z., 2002. Should strategic market management replace strategic management as the planning model for restaurant decision making? J. Foodserv. Bus. Res. 5 (1), 23–43.

54. Reynolds, D., Merritt, E.A., Pinckney, S., 2005. Understanding menu psychology: an empirical investigation of menu design and customer response. Int. J. Hosp. Tourism Adm. 6 (1), 1–9.

55. Ryu, K., Han, H., 2011. New or repeat customers: how does physical environment influence their restaurant experience? Int. J. Hosp. Manag. 30 (3), 599–611.

56. Ryu, K., Jang, S.S., Sanchez, A., 2003. Forecasting methods and seasonal adjustment for a university foodservice operation. J. Foodserv. Bus. Res. 6 (2), 17–34.

57. Sanlier, N., Comert, M., Durlu-Ozkaya, F., 2010. Hygiene perception: condition of hotel kitchen staffs in Ankara, Turkey. J. Food Saf. 30, 415–431.

58. Sarvan, F., Arıcı, E.D., Ozen, J., Ozdemir, B., Icigen, E.T., 2003. Ten schools of strategic management: the integrative framework of configuration school. Akdeniz Univ. Fac. Econ. Adm. J. 6 (3), 73–122.

59. Shoemaker, S., 1994. A proposal to improve the overall price value perception of a product line. J. Restaur. Foodserv. Mark. 1 (1), 89–101.

60. Shoemaker, S., Dawson, M., Johnson, W., 2005. How to increase menu prices without alienating your customers. Int. J. Contemp. Hosp. Manag. 17 (7), 553–568.

61. Taylor, J., Brown, D.M., 2007. Menu analysis: a review of techniques and approaches. FUI Hosp. Rev. 25 (2), 74–82.

62. Taylor, J., Reynolds, D., Brown, D.M., 2009. Multi-factor menu analysis using data envelopment analysis. Int. J. Contemp. Hosp. Manag. 21 (2), 213–225.

63. Vaughn, P., Raab, C., Nelson, K.B., 2010. The application of activity-based costing to a support kitchen in a Las Vegas casino. Int. J. Contemp. Hosp. Manag. 22 (7), 1033–1047.

64. Wansink, B., Painter, J., Van Ittersum, K., 2001. Descriptive menu labels' effect on sales. Cornell Hotel Restaur. Adm. Q. 42 (6), 68–72. Wansink, B., Van Ittersum, K., Painter, J.E., 2005. How descriptive food names bias sensory perceptions in restaurants. Food Qual. Prefer. 16 (5), 393–400.

65. Worsfold, D., 2006. Consumer information on hygiene inspections of food premises. J. Food Serv. 17 (1), 23–31.

66. Yang, S.S., 2012. Eye movements on restaurant menus: a revisitation on gaze motion and consumer scanpath. Int. J. Hosp. Manag. 31 (3), 1021–1029.

67. Yang, S.S., Kimes, S.E., Sessarego, M.M., 2009. Menu price presentation influences on consumer purchase behavior in restaurants. Int. J. Hosp. Manag. 28, 157–160.

2

CHAPTER

THE IMPACT OF RESTAURANTS' GREEN SUPPLY CHAIN PRACTICES ON FIRM PERFORMANCE

Jun-Zhi Chiu [1] and Chao-Chen Hsieh [2,*]

[1]*Graduate Institute of Hospitality Management, National Kaohsiung University of Hospitality and Tourism, No. 1, Songhe Rd., Xiaogang Dist., Kaohsiung City 81271, Taiwan*

[2]*Department of International Business, Chang Jung Christian University, 1 Changda Rd., Gueiren District, Tainan City 71101, Taiwan*

ABSTRACT

This study investigated crucial green supply chain management (GSCM) practice dimensions and firm performance based on restaurants firms in Taiwan. On the basis of a factor analysis, four green supply chain management dimensions were identified: corporate environment policy, green packing, green product, and economic transport. This study investigated crucial GSCM practice dimensions (including corporate environment policy, packaging waste, economic transport, and product recycling), green capability and organizational performance. The results shown: first, green practices in restaurants in Taiwan have an indirect effect on firm performance through green capability; second, when the ability of suppliers of green and green capability is at a higher degree, it will contribute to organizational performance, namely environmental and economic performance. Finally, green practices could be a key driver of green capability and it should be a priority in restaurants.

Keywords: green supply chain management; green capability; firm performance

1. INTRODUCTION

The hospitality industry is considered to be a tertiary Industry, but it does affect the environment [1], the tourism industry has been recognized as an important emitter of green house gases, particularly carbon dioxide. It is important to take action to reduce the related environmental impact on the environment in the hospitality industry. Therefore, with their share in rising environmental awareness, many companies have successively declared their environmental and social responsibility. Otherwise, firms should pay attention to their awareness and ability to respond due to the environmental impact in the food and beverage industry. Companies play an integral role to prevent environmental impact, for example they are responsible for environmental degradation caused by the production process. Hence, many NGOs have the information they require companies to publish the impact of the environment, and customers have begun to look for environmentally friendly products to purchase, and began to demand that the company has a green system [2]. Previous research has generally found that environmental practices have a positive impact on organizational performance in terms of operations, composition savings, competitiveness, firm image, stakeholder satisfaction and the financial performance of services. Skjoett-Larsen found that the green supply chain is one of the most important challenges of this century's European companies [3]. In the future, "green" or "environment", will become an important competitive factor in the SCM system [3,4]. The enterprise should be "green" as a competitive ability, representatives active strategy, replacing the relevant provisions of past environmental compliance more passive way [5], and that the enterprise will receive if improved environmental performance better competitiveness [6].

Past studies have shown that supply chain management can improve organizational performance. Supply chains and organizations can gain a competitive advantage by being the first to adopt environmental sustainability and implement GSCM practices [7,8]. While focusing on your suppliers' practices, consider the importance of your own eco-friendly strategies as the foundation of your operation's wherewithal. In other words, greening the supply chain means using a systematic approach to choose suppliers whose products/services are proven to be more sustainable than those of their competitors, and whose own green practices are more sustainable. Additionally, firms must rethink business strategy and the development of green innovation practices to respond immediate needs for the market and environmental challenges [9]. In summary, the study investigated crucial GSCM dimensions and from capability view, explore green capability antecedents and organizational performance as a result of the relationship.

2. LITERATURE REVIEW AND HYPOTHESES DEVELOPMENT

2.1. Green Supply Chain

The supply chain consists of all parties that are involved in fulfilling a customer request, including the suppliers, transporters, warehouses, retailers and customers themselves [10]. Green supply chain management (GSCM) practices involve organizations assessing the environmental performance of their suppliers, requiring suppliers to undertake measures that ensure the environmental quality of their products, and evaluating the cost of waste in their operating systems [11]. GSCM definition has ranged from green purchasing to integrated supply chain flowing from supplier, to manufacturer, to customer and reverse logistics [12].

Hervani *et al.* [13] explore GSCM activities including green design, purchasing, manufacturing/processing, production, marketing, recycling and material source. Srivastava describes GSCM as combining environmental thinking and SCM and defines it as including product design, material sourcing and selection, manufacturing processes, delivery of the final product to the consumer, and end-of-life management of the product after its useful life [14]. GSCM, advocating efficiency and synergy between partners, facilitates environmental performance, minimal waste and cost savings [15]. Comparing the definitions of SCM and GSCM, it is clearly shown that green supply chain involved not only integrating the manufacturing process and distribution to customers. GSCM also covers from the very beginning stage when the product is designed until the product is disposed. However, it requires a full integration and collaboration of player along the product life cycle.

Table 1. Construct definitions.

Construct	Definition
Corporate environment policy	Restaurant waste properly handled, using local produce and saving food miles concept. Achieving sustainable food, furniture and environmentally friendly products and the acquisition of the green mark, environmentally friendly and designed to build energy-saving green building materials [16]
Packaging waste	The management of packaging waste is therefore an important financial and environmental issue for restaurants in developing countries and economies in transition: it encourages continuous improvement in products and services; serves as a strategic tool for cutting costs; the decrease in processed and packaged foods; improves your public image; and promotes sustainable development
Economic transport	Based on environmental sustainability, through the use of an energy-efficient transport system and to enhance the transport efficiency of energy use, reducing transport vehicle mileage, *etc.*, restaurant food achieve a low-carbon transport activities
Product recycling	Recycling all resources that can be recycled, for example: plastic, paper, aluminum cans, glass bottles, cardboard boxes, and even cooking oil, reduce the use of a product that is lost. Reward guests for dining with recyclable mugs and containers
Green capability	Green capability relates to the manufacturing plant's capabilities to more efficiently produce and improve firm's strategic position in the market, whether domestic or international and as signals of a firm's competitive aggressiveness, capabilities, and market position
Economic performance	Economic performance relates to the manufacturing plant's ability to reduce costs associated with purchased materials, energy consumption, waste treatment, waste discharge, and fines for environmental accidents
Environmental performance	Environmental performance relates to the ability of manufacturing plants to reduce air emissions, effluent waste, and solid wastes and the ability to decrease consumption of hazardous and toxic materials

2.2. Motivations for the Adoption of Green Practices in Companies

Definitions of the constructs incorporated in the model are provided in Table 1. Generally, GSCM practices are the focal constructs in the theorized model with internal environmental management and green information systems as antecedents and green capability, economic performance and environmental performance, as consequences.

There are several researches that disclose that supply chains and organizations can gain green capability by being the first to adopt environmental sustainability and implement GSCM practices. For example, Diabat and Govindan studied various drivers that affect the implementation of green supply chains and found that green design positively influences the performance of green supply chains [17]. The aim of eco-design is the reduction of a product's environmental impact without creating a negative trade-off with other design criteria, such as costs and functionality. Seuring cautions that transaction costs associated with interactions among supply chain partners must be considered as the partners work to improve the environmental sustainability of the supply chain [18]. Both Pereira-Moliner and Molina-Azorín *et al.* [19,20] found hotels that adopted more environmental practices achieved better firm performance than those adopting fewer.

Bansal and Roth acknowledged the pursuit of competitive advantage as a focal firm's environmental policy efforts, which contributes to a firm's environmental management activity [21]. Firms implement environmental policy activities in an effort to create efficiencies and to obtain stakeholder approval and, thus, gain competitive advantage and realize superior returns [22]. In this study, we are particularly interested in examining the competitive drivers of environmental policy activities. Porter explains, "a signal is any action by a competitor that provides a direct or indirect indication of its intentions, motives, goals, or internal situation" (p. 75) [23]. In essence, signals convey information about product quality, firm reputation, or the competitive intentions of rivals to market participants. Hence, green capability and reactions signal to the market and to stakeholders the value that a firm is creating for them

[24]. Green capability serves as signals of a firm's competitive aggressiveness, capabilities, and market position [25].

In accordance with prior research, the following hypothesis is proposed:

H1: Corporate environment policy is positively related to restaurant's green capability.

Environmental packing procurement is an increasingly important issue and decision-making tool for small and. It can be ignored by an enterprise only at peril to its long-term survival. The EU Directive on Packaging and Packaging Waste 94/62/EC seeks to reduce the impact of packaging and packaging waste on the environment. This does not apply directly to procurement services but will affect companies in our supply chain. The Welsh Assembly Government's long term strategy for the environment of Wales, setting the strategic direction for the next 20 years including the need for more efficient materials consumption and waste management. Procurement Services has a role to play in contributing to the desired outcomes of the strategy, such as through encouraging our supply chain to design products (and packaging) that require less resources in their production, use and end of life, that create minimal waste and are easily reused or recycled. In the research of Hayami *et al.* (2015) to encouraging suppliers to reduce waste can lead to internal green product, process and managerial innovations and can enhance competitive advantage [26]. If this is so, then the main driving forces for implementation of environmental management include not only compliance with regulations and legislation but also cost savings. Therefore, the next hypothesis is proposed:

H2: Packaging waste is positively related to restaurant's green capability.

Zsidisin and Siferd pointed out suppliers should not only be evaluated, but that supplier development should also be part of green purchasing [27]. Working with sustainable supplier development can also be seen as a way to reduce the environmental risk in supplier relationships, something that should also have positive effects on operational costs and overall competitive advantage [28]. In addition, supplier selection and development (SSD) are crucial processes for successful GSCM. If problems surface regarding waste levels, transportation of hazardous materials or carbon emissions, suppliers may not be considered for future contracts. Firms should enhance green procurement and stimulate their transport and logistics suppliers in developing collaborative initiatives. Through collaboration with other suppliers the size of the counterpart as well as enhanced buyer competence should reduce and share the risk associated with specific resource investment in the area of green logistics which was support by Large *et al.* [29] who suggest that green collaboration with a supplier supposedly influences the environmental performance of the supplier. Therefore, the next hypothesis is proposed:

H3: Economic transport is positively related to restaurant's green capability.

The European Union, for example, passed the WEEE (Waste Electrical and Electronic Equipment) and RoHS (Restriction of Hazardous Substance) regulations, unequivocally proscribing the use of six hazardous substances in products for sale in the European market, and requiring compliance with the relevant product recycling laws. Moreover, the importance of GSCM activities such as green design, purchasing, manufacturing/processing, production, marketing, recycling and material source, has been explored in the green supply chain literature [13]. Diabat and Govindan find reducing energy consumption, reusing and recycling materials and packaging are influenced the implementation of GSCM [17]. Li also demonstrated recycling products processing ability was strongly balancing to other advanced management practices, and contributed to improving environmental performance [30]. Based on the findings listed above, the following hypothesis is proposed:

H4: Product recycling is positively related to restaurant's green capability.

2.3. Impact of the Adoption of Green Practices on Firm Performance

Hoejmose *et al.* [31] remark that restaurants experience a high level of stakeholder scrutiny while using green practices to achieve green capability. Consequently, when a firm adopts such practices, it can improve its image, attract new, "green" consumers, further satisfy existing customers and thus ensure themselves a better overall position in the market [32]. Furthermore, firm usually implement these steps to achieve consumption economic savings. Green practices do not require significant financial investment but may provide immediate economic benefits [33]. For example, through the application of more efficiently water and energy consumption and waste recycling methods and the subsequent bring reduction in total operational costs.

Based on the findings listed above, the following hypothesis is proposed:

H5: Green capability is positively related to a restaurant's economic performance

Enterprises to implement green supply chain management to increase efficiency and promote business relations between partners, in order to improve environmental performance, waste minimization and cost [34,35]. Chen believes that the implementation of environmental management strategies of the enterprise, not only to avoid penalties and regulations, but also to enhance their green image [36]. Our focus is on the sustainability incorporates the concepts of environmental performance has moved from the organization level to the supply chain level [37,38]. Seuring (2004) describes green capability as the managerial integration of material and information flows throughout the supply chain to satisfy the demand of customers for green products and services produced by green processes [39]. Supply chains strive to maintain internal health and environmental sustainability using the capability to correct based on information from the external environment [40].

Therefore, the next hypothesis is proposed:

H6: Green capability is positively related to restaurant's environmental performance.

3. MATERIALS AND METHODS

3.1. Data Collection

Data were obtained from on-site employee and shopkeeper in the samples of this study. Overall, 130 respondents from Taiwan completed the survey. The survey questionnaires included three sections. The survey questionnaires included three sections. The first section investigated the demographic profile of respondents and the second section included twenty-one statements/variables about the green supply chain practice. The twenty-one statements represented four green supply chain practice including "corporate environment policy," "packaging waste," "economic transport", and "product recycling". This section based on the comments from senior chain restaurants staffs are asked to clarify the belief statements and comment on their appropriateness for investigating green capability in the context of Taiwan. Respondents were asked to express the extent to which they disagreed or agreed with the statements. The second section of the questionnaire included sixteen questionnaires of firm performance including "economic performance," "green capability," and "environmental performance". The third section of the questionnaires was designed based on the literature concerned with firm performance. The refined types of green supply chain practice were perceived as the predictor variables, while the three types of firm performance were regarded as the criterion variables.

3.2. Participants

The sample for this study consisted of all managerial employees of the chain restaurants were set up in Taiwan with around 170 shopkeeper in Taiwan Association of F&B Societies. Of the 135 questionnaires distributed, 130 shopkeeper responded, yielding a response rate of 96.30%. Five of the returned questionnaires were excluded from the analysis because of missing pages. Of the total 130 subjects, 50.8% were female. The age of respondents ranged from 30 to 60 years. A significant percentage of respondents (31.5%) were in the 31-year-old to 40-year-old age group. In terms of education, 27.7% had completed high school. 55.45% had a bachelor's degree, and 1.5% had master degree. The age of respondents ranged from 20 to 60 years. A significant percentage of respondents (62.3%) were in the 31- to 50-year-old age group. Tenure in the organization ranged from 3 years to 9 years, with a mean of 6 years (45.4%). In terms of education, 55.4% had completed high school and 27.7% had a bachelor's degree.

3.3. Measures

Measures of the survey were initially developed based on input from a thorough literature review. Drafted measuring instruments were then assessed by three academics knowledgeable about the topic under investigation. The assessed instruments were further pre-tested with 20 senior executives working for the F&B industry in Taiwan. All executives who participated in this instrument development process were excluded from the subsequent mail survey. English was initially used to develop the survey questionnaire, which was subsequently translated into Chinese. The linguistic equivalence between the two versions was ensured by employing the back-translation technique.

3.3.1. Corporate Environment Policy

Four items each for corporate environment policy were adapted from Bagur-Femenias *et al.* (2013) relevant scales [41]. The items were coded on a five-point scale ranging from 1 = "strongly disagree" to 5 = "strongly agree".

3.3.2. Packaging Waste

To measure packaging waste, the scale proposed by Rao (2005) was used, Such as the optimization of processes to reduce solid waste [15]. The items were coded on a five-point scale ranging from 1 = "strongly disagree" to 5 = "strongly agree".

3.3.3. Economic Transport

As mentioned, this investigation adapted proposed Rao (2005) proposed items to capture the three major economic transpo rt s, such as change for more environmentally-friendly transportation [15]. The items were coded on a five-point scale ranging from 1 = "strongly disagree" to 5 = "strongly agree".

3.3.4. Product Recycling

Three items each for corporate environment policy were adapted from Rao (2005) relevant scales [15]. The items were coded on a five-point scale ranging from 1 = "strongly disagree" to 5 = "strongly agree".

3.3.5. Green Capability

The green capability of restaurants reflects their ability to manage their environmental impacts through provision of environmentally consci ous products, adopting cleaner production processes, integrating environmental issues into business routines, sourcing from environmentally responsible upper-tier suppliers, and communicating their environmental policy with partners [26,42]. A four-item scale is adopted from Lee and Klassen (2008) Wong *et al.* (2012) and Bagur-Femenias *et al.* (2013) [41,42,43]. We asked respondents to assess their firms on these items using a five-point Likert scale with 1strongly disagree and strongly agree.

3.3.6. Economic Performance

To measure economic performance, the scale proposed by Rao (2005) was used. Such as new market opportunities [15]. The items were coded on a five-point scale ranging from 1 = "strongly disagree" to 5 = "strongly agree".

3.3.7. Environmental Performance

The three-item environmental performance scale was based on a measure developed by Zhu *et al.* (2008) and Bagur-Femenias *et al.* (2013) [41,44]. Therefore, we tested the model using a perceptual measure of environmental performance in which each respondent rated his or her organization's environmental performance relative to that of other firms in the restaurants using three items. It was used to assess overall environmental performance. The responses were on a five-point Likert-type scale, ranging from 1 = "strongly disagree" to 5 = "strongly agree".

4. RESEARCH METHODOLOGY

4.1. Measurement Development

A descriptive analysis was given in Table 2 used to investigate the demographic information of the respondents, including type of operation and number of employees. A principal component analysis with a varimax rotation was used to test whether twenty-one statements/variables were loaded on five types of green supply chain practice and sixteen statements were loaded on three types of firm performance. The types of GSCM practice and firm performance were therefore refined, and the internal consistency of the refined orientations was examined using Cronbach's alpha reliability coefficients.

A partial least square structural equation model was constructed to test both the reliability and validity of measurement and the structural model. The measurement model represents the relationship between the manifest variables (measurable items) and the unobserved latent variables, while the structural model focused on the hypothesized relationship between the latent variables [45]. A first order reflective model was employed in this study since the latent construct existed independent of the measures used; the variation in the construct caused variation in the item measures; this variation did not cause a variation in the construct; the items shared a common theme; the items were interchangeable so that adding or dropping an item did not change the conceptual domain of the construct; and the items should have had high positive intercorrelations in reflective models [46,47].

However, the PLS approach was employed in this study since PLS: (1) places less premium on explaining the covariance of all item measures, (2) avoids a negative impact due to errors in modeling or item usage, (3) values soft distributional assumptions, (4) sees the research as not simply exploratory in nature, but interactive, (5) has formative measurement items (even if the formative modelwas not used), (6) requires flexibility in modeling higher order Molar and Molecular models, (7) is interested in obtaining determinate scores/indices that are predictive (8) has high model complexity (9) faces a relatively smaller sample size, (10) is less concerned with the accuracy of parameter estimation or does not hold a belief in the notion of an underlying covariance based latent variable generating mechanism, (11) wants to shift the perspective of a "True" Model towards a Prediction Focus, and values ease of model specification and model interpretation [48]. Fornell and Bookstein stated that PLS is a powerful method of analysis because of its minimal demands on measurement scales, sample size, and residual distributions [49]. PLS avoids the two serious problems of inadmissible solutions and factor indeterminacy [49]. Many environmental based researchers have used the PLS method to analyze different relationships [50,51]. Smart PLS software was employed as the analysis tool in this research to test the proposed model. The results are shown below [52].

4.2. Measurement Model

To assess how well the given indicators explained the construct, testing of the measurement model was crucial. Therefore, the factor loadings, the composite reliability, the average variance extracted (AVE), as well as Cronbach's alpha, communality, convergent validity, and discriminant validity were used to test the measurement model. In essence, the reliability and internal consistency of the model were tested. According to the results, all the scores were above or near the threshold values. AVE scores for each construct ranged from 0.62 to 0.73, which exceeded the suggested threshold value of 0.5 [53]. The composite reliabilities were all well above the recommended threshold value of 0.8. Cronbach's alpha showed as above 0.76 the recommended threshold value. Convergent validity was used for assessing each indicator's relationship with the connected construct. The analysis of convergent validity is shown under factor loadings in Table 2, which shows how well each indicator was correlated with the construct that it is connected to. According to the table, factor loadings exceeded the value of 0.5 suggested by Hair *et al.* [45]. Examining the degree of divergence of each construct from each other is known as discriminant validity analysis. For this analysis, the square root of the AVE scores for each construct was compared with their correlation with the other constructs. It is recommended that the square root of AVE scores should be greater than their correlation with the other constructs [54].

Table 2. Reliability of the measurement model.

	Loadings	Composite Reliability	Average Variance Extracted	Cronbach's Alpha	Communality
Corporate environment policy	0.78 0.89 0.75 0.78	0.874	0.635	0.803	0.635
Packaging waste	0.88 0.85 0.81 0.66	0.869	0.629	0.791	0.629
Economic transport	0.89 0.88 0.71	0.820	0.699	0.767	0.699
Product recycling	0.92 0.88 0.77	0.892	0.736	0.811	0.736
Green capability	0.83 0.79 0.79 0.73	0.917	0.736	0.879	0.736
Economic performance	0.83 0.81 0.79 0.72 0.60	0.905	0.659	0.865	0.659
Environmental performance	0.90 0.89 0.52	0.905	0.687	0.763	0.687

The proposed hypothesized relationships, shown in Figure 1 were investigated in relation to the size of the firms. The robustness and reliability of the measurement model were also investigated by using average variance extracted (AVE), composite reliability and Cronbach's alpha. There was no issue with the reliability of the measurement model in each sector group. The results demonstrated that the hypothesized relationships.

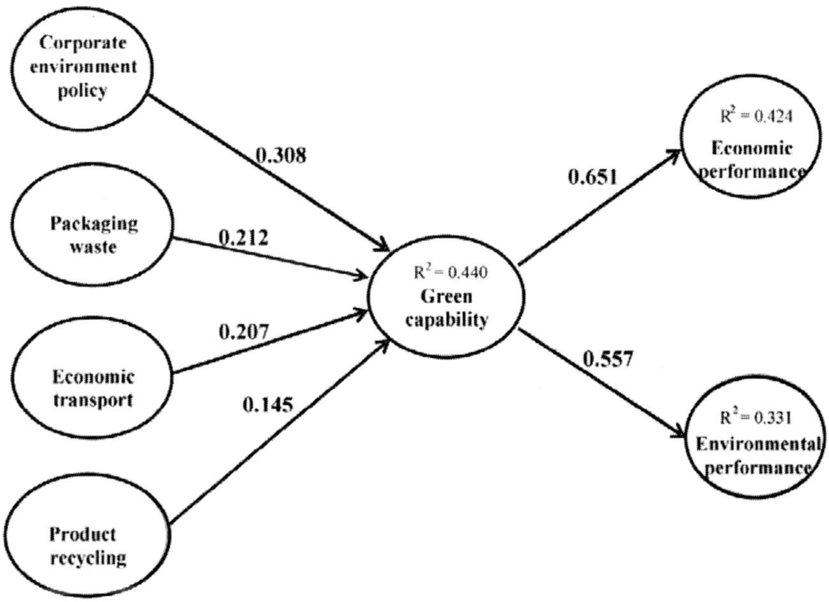

Figure 1. Hypothesized relationships among green supply chain management (GSCM) practices, green capability and green performance.

Table 3 reveals the correlation matrix of constructs and the square root of AVE scores in bold. According to the obtained result, off-diagonal results (correlation of constructs) were less than or equal to the bolded diagonal (square root of AVE scores) values. There was no problem with discriminant validity, which indicated the validity of the proposed measurement model. It was therefore concluded that the measurement model was reliable.

Table 3. Correlation of constructs and square root of AVE.

	GC	ECP	ENP	CEP	PW	PR	ET
Green capability(GC)	0.857						
Economic performance(ECP)	0.650	0.811					
Environmental performance(ENP)	0.557	0.458	0.828				
Corporate environment policy(CEP)	0.518	0.357	0.443	0.796			
Packaging waste(PW)	0.535	0.352	0.626	0.452	0.793		
Product recycling(PR)	0.436	0.471	0.491	0.323	0.432	0.857	
Economic transport(ET)	0.501	0.455	0.459	0.324	0.586	0.483	0.836

Diagonals show the square root of AVEs.

4.3. Structural Model

A structural model was used to assess the hypothesized relationship of the constructs, the coefficient of determination (R^2), the path coefficient and their corresponding significance scores. To assess the statistical significance between latent variables (constructs), traditional parametric tests were inappropriate in PLS [54]. Therefore, bootstrapping as a non-parametric test was used to test the hypothesized relationship between constructs. In order to do this, 200 bootstrap samples were established by resampling with replacement from the original samples. Table 4 shows that the R^2 value of green capability was 44%, the economic performance was 42.3%, and the environmental performance was 31% of the total variance.

Table 4. R square values.

Constructs	R^2
Green capability	0.440
Economic performance	0.423
Environmental performance	0.310

In order to assess overall performance of the model as well as the measurement and the structural models, a global criterion of goodness of fit index (GoF index) as developed by Tenenhaus *et al.* [55] was employed. The GoF index provides a single measure for the overall prediction performance of the model by taking the geometric mean of the average communality index and the average R^2 value $\sqrt{\text{Average} \times \text{Communality Average} \times R^2}$. The GoF index of the model was 0.542 which indicated that the model was able to take into account 54.2% of the achievable fit.

Table 5 shows the path coefficient results as well as their t-values which were obtained by using the bootstrapping re-sampling technique in Smart PLS. The path coefficients and their associated *t*-values demonstrated the direction and strength of each hypothesized relationship. Based on obtained results, there was a positive significant relationship between CEP and GC. It was found that the hypothesized relationship was statistically significant at the $p < 0.01$ level with a β = 0.308 path coefficient (CEP→GC; $\beta = 0.308, p < 0.01$). In addition, PW→ GC ($\beta = 0.212, p < 0.1$), ET→ GC ($\beta = 0.207, p < 0.1$), there was a weak positive significant relationship to the, PR→GC; $\beta = 0.145$, GC→EcP ($\beta = 0.651, p < 0.001$), GC→EnP ($\beta = 0.557, p < 0.001$). In conclusions, except H_4 each of the five hypothesis were support at the 0.1 level.

Table 5. Summary of testing results.

Relationship	Hypothesis	Coefficient	*t*-Value	Hypothesis Test Result
CEP→ GC	H_1	0.308	2.994 **	supported
PW→ GC	H_2	0.212	2.099 *	supported
ET→ GC	H_3	0.207	2.024 *	supported
PR→ GC	H_4	0.145	1.550	not supported
GC→ EcP	H_5	0.651	11.896 ***	supported
GC→ EnP	H_6	0.557	8.472 ***	supported

* $t > 1.96$ ** $t > 2.58$ *** $t > 3.3$.

5. DISCUSSION

Hypotheses test results are presented in Table 5. Except for H_4 each of the five hypothesis were supported at the 0.1 level. Although the product recycling has a weak direct effect on green capability, in contrast to previous research on the food and beverage industry, this study does not find a positive effect between product recycling and green capability. Thus, it is necessary to assess the effect of environmental practices as a whole, which could be difficult for restaurants because the attributes of management systems are generally informal.

There are three reasons that express why product recycling is a limited resource being used faster in chain restaurant: first, most commonly used restaurant products like plastic wrap, toothpicks, paper napkins and towel, and straws can be purchased produced with recycled, biodegradable, tree-free, or organic components [56]. The cost, however, is what would make a restaurant think twice. Second, simply using a recycling program or changing standard operating procedures to lessen the amount of time certain appliances are left on will not affect a customer's perception of a restaurant. According to a study conducted by the NPD Group, despite the restaurant industry's effort to make operations more environmentally friendly, consumers do not view this sector as having made any progress with regards to going green [57]. This lack of awareness is directly contributing to the customer perception of the industry. In order to overcome this situation, it is important for third party certification organization to arise and make the efforts of the industry more relatable to customers. Third, Spanish regulations establish minimum sanitary controls and food traceability on operations that are mandatory; however, restaurant managers are free to determine the use of environmental practices within operations as a part of the company's strategy. Thus, it appears that restaurants reacted to this pressure by implementing green operational practices such as conserving water, conserving energy, recycling hazardous product waste or using ecological food. This situation is the same in Taiwan. Previous studies (Ilomaki and Melanen 2001; Jacobs *et al.*2010) indicate a lack of interest among small service companies with regard to the implementation of green practices because they could not perceive a direct effect on business performance. Our results reinforce the belief that there is not always a direct effect between product recycling and green capability [58,59].

6. CONCLUSIONS

This study has examined the GSCM in a Taiwanese restaurant context. The study's main findings are summarized below. First, green practices in restaurants in Taiwan have an indirect effect on firm performance through green capability. The effect on green capability linked to the implementation of green practices has been widely studied [19,31]. This study reinforces previous findings and provides new evidence for the restaurant sector. Thus, green practices can improve capability both internally and externally. Externally, the restaurant's reputation improves the customers' perceptions of it [60,61,62,63], so green practices could be a good way to maintain positioning in the market during times of crisis [64]. Furthermore, the adoption of these measures can be used to attract a new type of customer to the restaurant by offering new menu selections according to their preferences and concerns thereby increasing sales more than the competition does [65,66]. Second, when the ability of suppliers of green and green capability is higher degree, it will contribute to organizational performance, namely environmental and economic performance which supported by Bowen *et al.* and Tseng *et al.*[34,35]. Restaurants implement GSCM practices to increase efficiency and promote business relations between partners, in order to improve environmental performance, waste minimization and cost savings. Additionally, when a higher degree of market-oriented firms faces the challenge, the higher the green capability firms have. This means that restaurants that own GSCM capability in the face of market-oriented firms will enhance their ability to make the ring environmental performance and economic performance has grown. Thus, the present study supports Zhu *et al.* [44]. when companies face external green market proposed to change the traditional mode of production and services for the new type of green management mode is available to develop green performance.

Finally, it appears that the improvement in terms of green capability overcomes the investment and additional costs related to the implementation and monitoring of green practices. Thus, this finding reinforces the results of previous studies [31,61]. Making a change to green practices could be a key driver of green capability and it should be a priority in restaurants. Further, the finding light on restaurants provide managers with insights into the characteristics of restaurants related supplier which are useful to them for formulating plans to benchmark and set targets for GSCM capability and performance improvement.

7. MANAGERIAL PRACTICES

With regard to managerial practices, it is possible to draw two interesting conclusions from this study that may be of direct interest to practitioners working in the sector. First, sustainability is increasingly valued in times of recession. Green practices in restaurants in Taiwan have a direct effect on firm performance through green capability and can efficiency improvement and competitive improvement that restaurants can develop to survive economic downturns and achieve better positioning in the market. Second, practitioners must consider green practices to be a way to improve competitiveness. Internally, senior managers should consider that the adoption of certain green practices improves employee performance and achieves cost reduction in consumption. Thus, the application and communication of these policies increases the motivation of workers because they feel that they are part of an environmentally sustainable project. Ultimately, customers receive better quality service, which may positively affect the profit and loss statement. Externally, green practices can be used to build and reinforce the brand. It is clear that being a green restaurant helps to cement customer loyalty to the brand and generates new customer interest in the restaurant's products. Thus, attaching the brand to a "green-friendly concept" may help practitioners to better satisfy customer needs.

8. LIMITATIONS AND AREAS FOR FUTURE STUDY

Several limitations of this study that need further attention in the future should be noted. First, because all of the measures used are self-reported, common method variance due to single-source bias is a problem and may have inflated the magnitude of the relationships found. Second, the relatively small sample used for this study may limit the generalizability of the results. Future research is also needs to examine the influence of national or restaurants organizational culture on the proposed relationships, and other mediators, such as mark orientation, should also be explored.

AUTHOR CONTRIBUTIONS

Jun-Zhi Chiu designed this study, collected and processed the data. Chao-Chen Hsieh provided some significant advice and checked through the whole paper. The first author wrote the paper and coauthors read and approved this version.

REFERENCES

1. Kasim, A. Towards a wider adoption of environmental responsibility in the hotel sector. *Int. J. Hosp. Tour. Admin.* 2007, *8*, 5–49.

2. Han, H.; Hsu, L.; Sheu, C. Application of the theory of planned behavior to green hotel choice: Testing the effect of environmental friendly activities. *Tour. Manag.* 2011, *31*, 325–334.

3. Skjoett-Larsen, T. European logistics beyond 2000. *Int. J. Distrib. Logist. Manag.* 2000, *30*, 377–387.

4. Abukhader, S.M.; Jönson, G. Logistics and the environment: Is it and established subject? *Int. J. Distrib. Logist. Res. Appl.* 2004, *7*, 137–149.

5. Van Hoek, R.I. From reversed logistics to green supply chains. *Supply Chain Manag.* 1999, *4*, 129–124.

6. Bacallan, J.J. Greening the supply chain. *Bus. Environ.* 2000, *6*, 11–12.

7. Barratt, M.; Oke, A. Antecedents of supply chain visibility in retail supply chains: a resource-based theory perspective. *J. Oper. Manag.* 2007, *25*, 1217–1233.

8. Sen, S. Linking green supply chain management and shareholder value creation. *IUP J. Supply Chain Manag.* 2009, *7*, 95–109.

9. Jaworski, B.J.; Kohli, A.K. Market Orientation: Antecedents and Consequences. *J. Mark.* 1993, *57*, 53–70.

10. Cox, A. Power, value and supply chain management. *Supply Chain Manag.* 1999, *4*, 167–175.

11. Handfield, R.; Walton, S.; Sroufe, R.; Melnyk, S. Applying environmental criteria to supplier assessment: A study in the application of the analytical hierarchy process. *Eur. J. Oper. Res.* 2002, *141*, 70–87.

12. Zhu, Q.; Sarkis, J. Relationships between operational practices and performance among early adopters of green supply chain management practices in Chinese manufacturing enterprises. *J. Oper. Manag.* 2004, *22*, 265–289.

13. Hervani, A.A.; Helms, M.M.; Sarkis, J. Performance measurement for green supply chain management.*Benchmarking Int. J.* 2005, *12*, 330–353.

14. Srivastava, S.K. Green supply-chain management: A state-of-the-art literature review. *Int. J. Manag. Rev.*2007, *9*, 53–80.

15. Rao, P.; Holt, D. Do green supply chains lead to competitiveness and economic performance? *Int. J. Oper. Prod. Manag.* 2005, *25*, 898–916.

16. Lorenzini, B. The green restaurant, part II: Systems and service. *Restaur. Inst.* 1994, *104*, 119–136.

17. Diabat, A.; Govindan, K. An analysis of the drivers affecting the implementation of green supply chain management. *Resour. Conserv. Recycl.* 2011, *55*, 659–667.

18. Seuring, S. Green supply chain costing: joint cost management in polyester linings supply chain. *Greener Manag. Int.* 2001, *33*, 71–80.

19. Pereira-Moliner, J.; Claver-Cortés, E.; Molina-Azorín, J.F.; Tarí, J.J. Quality management, environmental management and firm performance: Direct and mediating effects in the hotel industry. *J. Clean. Prod.* 2012,*37*, 82–92.

20. Molina-Azorín, J.F.; Claver-Cortes, E.; Pereira-Moliner, J.; Tarí, J.J. Environmental practices and firm performance: An empirical analysis in the Spanish hotel industry. *J. Clean. Prod.* 2009, *17*, 516–524.

21. Bansal, P.; Roth, K. Why companies go green: A model of ecological responsiveness. *Acad. Manag. J.* 2000,*43*, 717–736.

22. Montabon, F.; Sroufe, R.; Narasimhan, R. An examination of corporate reporting, environmental management practices and firm performance. *J. Oper. Manag.* 2007, *25*, 998–1014.

23. Porter, M.E. *Competitive Strategy*; The Free Press: New York, NY, USA, 1980.

24. Basdeo, D.K.; Smith, K.G.; Grimm, C.M.; Rindova, V.P.; Derfus, P.J. The impact of market actions on firm reputation. *Strateg. Manag. J.* 2006, *27*, 1205–1219.

25. Grimm, C.M.; Lee, H.; Smith, K.G. *Strategy as Action: Competitive Dynamics and Competitive Advantage*; Oxford University Press: Oxford, UK, 2005.

26. Hayami, H.; Nakamura, M.; Nakamura, A. Economic performance and supply chains: The impact of upstream firms' waste output on downstream firms' performance in Japan. *Int. J. Oper. Econ.* 2015, *160*, 47–65.

27. Zsidisin, G.A.; Siferd, S.P. Environmental purchasing: A framework for theory development. *Eur. J. Purch. Supply Managet.* 2001, *7*, 61–73.

28. Foerstl, K.; Reuter, C.; Hartmann, E.; Blome, C. Managing supplier sustainability risks in a dynamically changing environment—Sustainable supplier management in the chemical industry. *J. Purch. Supply Managet.* 2010, *16*, 118–130.

29. Large, R.O.; Thomsen, C. G. Drivers of green supply management performance: Evidence from Germany. *J. Purch. Supply Manag.* 2011, *17*, 176–184.

30. Li, Y. Research on the performance measurement of green supply chain management in China. *J. Sustain. Dev.* 2011, *4*, 101–107.

31. Hoejmose, S.; Brammer, S.; Millington, A. An empirical examination of the relationship between business strategy and socially responsible supply chain management. *Int. J. Oper. Prod. Manag.* 2013, *33*, 589–621.

32. Hillary, R. Environmental management systems and the smaller enterprise. *J. Clean. Prod.* 2004, *12*, 561–569.

33. Zeng, S.X.; Meng, X.H.; Yin, H.T.; Tam, C.M.; Sun, L. Impact of cleaner production on business performance. *J. Clean. Prod.* 2010, *18*, 975–983.

34. Bowen, F.E.; Cousins, P.D.; Lamming, R.C.; Farukt, A.C. Horses for courses: Explaining the gap between the theory and practice of green supply. *Greener Manag. Int.* 2001, *35*, 41–60.

35. Tseng, M.L.; Chiu, A.S.F. Evaluating firm's green supply chain management in linguistic preferences. *J. Clean. Prod.* 2010, *40*, 22–31.

36. Chen, Y.S. The driver of green innovation and green image—Green core competence. *J. Bus. Ethics* 2008, *81*, 531–543.

37. Linton, J.D.; Klassen, R.; Jayaraman, V. Sustainable supply chains: an introduction. *J. Oper. Manag.* 2007, *25*, 1075–1082.

38. Preuss, L.L. Green light for greener supply. *Bus. Ethics Eur. Rev.* 2002, *4*, 308–317.

39. Seuring, S. Industrial ecology, life cycles, supply chains: Differences and interrelations. *Bus. Strategy Environ.* 2004, *13*, 306–319.

40. Vachon, S. Green supply chain practices and the selection of environmental technologies. *Int. J. Prod. Res.*2007, *45*, 4357–4379.

41. Bagur-Femenias, L.; Llach, J.; Alonso-Almeida, M.M. Is the adoption of environmental practice a strategic decision for small service companies? An empirical approach. *Manag. Decis.* 2013, *51*, 41–62.

42. Wong, C.W.Y.; Lai, K.-H.; Shang, K.-C.; Lu, C.-S.; Leung, T.K.P. Green operations and the moderating role of environmental management capability of suppliers on manufacturing firm performance. *Int. J. Prod. Econ.* 2012, *140*, 283–294.

43. Lee, S.-Y.; Klassen, R.D. Drivers and enablers that foster environmental management capability in small-and medium-sized suppliers in supply chains. *Prod. Oper. Manag.* 2008, *17*, 573–586.

44. Zhu, Q.; Sarkis, J.; Lai, K. Confirmation of a measurement model for green supply chain management practices implementation. *Int. J. Prod. Econ.* 2008, *111*, 261–273.

45. Hair, J.F.; Anderson, R.E.; Tatham, R.L.; Black, W.C. *Multivariate Data Analysis*, 5th ed.; Prentice-Hall International, Inc.: Upper Seddle River, NJ, USA, 1998.

46. Edwards, J.; Bagozzi, R. On the nature and direction of relationships between constructs and measures.*Psychol. Methods* 2000, *5*, 155–174.

47. Diamantopoulos, A.; Siguaw, J.A. Formative *versus* reflective indicators in organizational measure development: A comparison and empirical illustration. *Br. J. Manag.* 2006, *17*, 263–282.

48. Chin, W.W. How to Write Up and Report PLS Analyses. In *Handbook of Partial Least Squares*; Esposito Vinzi, V., Chin, W.W., Henseler, J., Wang, H., Eds.; Springer-Verlag: Heidelberg, Germany, 2010; pp. 655–690.

49. Fornell, C.; Bookstein, F.L. Two structural equation models: Lisrel and PLS appliead to consumer exit-voice theory. *J. Mark. Res.* 1982, *19*, 440–452.

50. Monevan, J.M.; Ortas, E. Corporate environmental and financial performance: A multivariate approach.*Ind. Manag. Data Syst.* 2010, *110*, 193–210.

51. Akamp, M.; Müller, M. Supplier management in developing countries. *J. Clean. Prod.* 2013, *56*, 54–62.

52. Ringle, C.M.; Wendle, S.; Will, S. *Smart PLS 2.0.*; University of Hamburg: Hamburg, Germany, 2005.

53. Fornell, C.; Larcker, D.F. Evaluating structural equation models with unobservable variables and measurement error. *J. Mark. Res.* 1981, *18*, 39–50.

54. Chin, W.W. The partial least squares approach to structural equation modeling. *Mod. Bus. Res.* 1998, *295*, 295–336.

55. Tenenhaus, M.; Amato, S.; Esposito Vinzi, V. A Global Goodness-of-Fit Index for PLS Structural Equation Modelling. 2004. Available online: http://www.old.sis-statistica.org/files/pdf/atti/RSBa2004p739-742.pdf(accessed on 4 January 2016).

56. Green Home. 2008. Avaiable online: http://www.greenhome.com/business/green-food-service-supplies.html (accessed on 1 January 2016).

57. Glazer, F. NPD: Consumers perceive industry to be behind on green efforts. *Nation's Restaur. News* 2008, *14*, 50.

58. Ilomaki, M.; Melanen, M. Waste minimisation in small and medium sized enterprises- do environmental management systems help? *J. Clean. Prod.* 2001, *9*, 209–217.

59. Jacobs, B.W.; Singhal, V.R.; Subramanianet, R. An empirical investigation of environmental performance and the market value of the firm. *J. Oper. Manag.* 2010, *8*, 430–441.

60. Susskind, A.M. I told you so! Restaurant customers' word-of-mouth communication patterns. *Cornell Hotel Restaur. Adm. Q.* 2002, *43*, 75–85.

61. Susskind, A.M. Guest service management and processes in restaurants: what we have learned in fifty years. *Cornell Hosp. Q.* 2010, *51*, 479–482.

62. Longart, P. What drives word-of-mouth in restaurants? *Int. J. Cont. Hosp. Manag.* 2010, *22*, 121–128.

63. Hyun, S.S.; Kim, W. Dimensions of brand equity in the chain restaurant industry. *Cornell Hosp. Q.* 2011, *52*, 429–437.

64. Alonso-Almeida, M.M.; Bremser, K. Strategic responses of the Spanish hospitality sector to the financial crisis. *Int. J. Hosp. Manag.* 2013, *32*, 141–148.

65. Hu, H.; Parsa, H.G.; Self, J. The dynamics of green restaurant patronage. *Cornell Hosp. Q.* 2010, *51*, 344–362.

66. Jang, Y.J.; Kimb, W.G.; Bonn, M.A. Generation Y consumers' selection attributes and behavioral intentions concerning green restaurants. *Int. J. Hosp. Manag.* 2011, *30*, 803–811.

3

CHAPTER

EXPLORATORY ANALYSIS OF FAST-FOOD CHAIN RESTAURANT MENUS BEFORE AND AFTER IMPLEMENTATION OF LOCAL CALORIE-LABELING POLICIES, 2005–2011

Alexa Namba, MPH, Amy Auchincloss, PhD, MPH, Beth L. Leonberg, MS, RD, CSP, FADA, LDN, and Margo G. Wootan, DSc

Amy H. Auchincloss, PhD, MPH, Drexel University School of Public Health, 1505 Race Street, Mail Stop 1033, Philadelphia, PA 19102.
Alexa Namba, School of Public Health, Drexel University, Philadelphia, Pennsylvania; Beth L. Leonberg, College of Nursing and Health Professions, Drexel University, Philadelphia, Pennsylvania; Margo G. Wootan, Center for Science in the Public Interest, Washington, DC

ABSTRACT

Introduction

Since 2008, several states and municipalities have implemented regulations requiring provision of nutrition information at chain restaurants to address obesity. Although early research into the effect of such labels on consumer decisions has shown mixed results, little information exists on the restaurant industry's response to labeling. The objective of this exploratory study was to evaluate the effect of menu labeling on fast-food menu offerings over 7 years, from 2005 through 2011.

Methods

Menus from 5 fast-food chains that had outlets in jurisdictions subject to menu-labeling laws (cases) were compared with menus from 4 fastfood chains operating in jurisdictions not requiring labeling (controls). A trend analysis assessed whether case restaurants improved the healthfulness of their menus relative to the control restaurants.

Results

Although the overall prevalence of "healthier" food options remained low, a noteworthy increase was seen after 2008 in locations with menulabeling laws relative to those without such laws. Healthier food options increased from 13% to 20% at case locations while remaining static at 8% at control locations (test for difference in the trend, $P = .02$). Since 2005, the average calories for an à la carte entrée remained moderately high (approximately 450 kilocalories), with less than 25% of all entrées and sides qualifying as healthier and no clear systematic differences in the trend between chain restaurants in case versus control areas ($P \geq .50$). Conclusion These findings suggest that menu labeling has thus far not affected the average nutritional content of fast-food menu items, but it may motivate restaurants to increase the availability of healthier options.

INTRODUCTION

Beginning with New York City in 2008, 18 states and localities have implemented regulations requiring chain restaurants to post calorie information on menus and menu boards (1). This approach to addressing obesity aims to assist customers in making informed, lower-calorie choices. Although it is too early to assess the full impact of menu labeling on consumer choice, research to date has shown mixed results. One large study found that menu labels may help some consumers make lower-calorie selections (2,3), while another small study found no difference in average calories purchased (4).

Menu-labeling laws have the potential to affect not only consumer behavior, but also to prompt industry changes. The prospect of negative consumer reaction to high-calorie menu offerings and increasing consumer demand for lower-calorie menu items may motivate the restaurant industry to reduce portion sizes, alter preparation, and add healthier items to their menus. Restaurants may be particularly sensitive to the nutritional quality of children's menu items. Public concern regarding the adverse health effects of childhood obesity has already prompted increased scrutiny of children's meals (5–7) and children's vulnerability to food marketing (8). In addition, a recent study examined whether calories in adult entrées at fast-food and sit-down chain restaurants in a county with nutrition labeling declined after labeling was implemented (9). The study found that, on average, calories for fast-food chain restaurants decreased by 19 kilocalories (kcal) (standard deviation [SD], 91) over the observation period. However, no trend study has included menus before 2008, the year nutrition-labeling regulations were implemented in New York City, or compared labeled menus with menus from chain restaurants operating in jurisdictions without nutrition-labeling laws.

This exploratory study used 7 years of data, from 2005 through 2011, from a small sample of fast-food chain restaurant companies located in jurisdictions with and without menu-labeling requirements. Our objective was to determine whether the nutritional quality of menus changed over this 7-year time period and if changes differed depending on whether companies were in jurisdictions requiring menu labeling.

METHODS

This case-control trends study examined 7 years of nutritional data, from 2005 through 2011, from 5 chain restaurant companies operating in areas requiring nutrition labeling on menus (cases) and 4 chain restaurant companies operating in areas not yet required to provide nutrition labeling (controls). This study did not include human subjects so was declared exempt by the Drexel University institutional review board.

Fast-food chain restaurants

Chain restaurants were eligible for inclusion in the study if they were ranked among the top 50 quick-service restaurants in 2010 (10), served entrées (were not coffee shops), and met criteria regarding location and data availability. Restaurant selection aimed to include large restaurant chains, given their larger contribution to Americans' diets as compared with smaller chains. By using the *Top-50 QSR* (quick service restaurant) ranking (10), restaurants were considered for inclusion beginning with 1 and moving to 50. Location criteria were that the restaurant chain company had 20 or more locations nationally, that case restaurant chains had outlets in areas where menu-labeling regulations had been implemented before 2011, and that control restaurant chains did not have outlets in jurisdictions requiring menu labeling. Data criteria were that the restaurant chain's website posted nutrition information for 6 or more years of the study period, that nutrition values were listed for 90% or more of menu items, and that values for calories, saturated fat, and sodium were provided (cholesterol and fiber were also provided by the eligible restaurants, so we were able to include them in the analysis). Few chain restaurants consistently posted complete nutrition data during the study period. Only 9 restaurant chains met study criteria and were included (Table 1). For purposes of this article, the 9 restaurant chains included in our study were de-identified and assigned a letter, A through I.

Nutrition values for menu items

Historic data came from an archive of Internet data collected from publicly accessible Web pages beginning in 1996 (11). The archive shows chain restaurant company websites exactly as they appeared in previous years. Dates were printed on each Internet menu; thus, we were able to verify the year the nutritional information represented. When possible, menus were collected from June of each calendar year.

Table 1. Characteristics of fast-food chain restaurants included in analysis, 2005–2011

Chain Restaurant[a]	Type of Food	Available Menu Years	Has Children's Menu	Average Number Outlets	Average Revenue[b] (million $)
Case chain restaurants	**NA**	**NA**	**NA**	**3,963**	**$4,074**
Restaurant A	Burger	2005–2011	Yes	3,649	$3,010.00
Restaurant B	Chicken	2006–2011	Yes	5,055	$4,700.00
Restaurant C	Fish	2005–2011	No	964	$700.00
Restaurant D	Burger	2005–2011	Yes	3,572	$3,619.90
Restaurant E	Burger	2005–2011	Yes	6,576	$8,340.00
Control chain restaurants[c]	**NA**	**NA**	**NA**	**829**	**$1,081.00**
Restaurant F	Chicken	2005–2011	No	484	$712.80
Restaurant G	Burger	2005–2011	Yes	424	$689.10
Restaurant H	Burger	2005–2008, 2010–2011	Yes	1,692	$1,695.00
Restaurant I	Burger	2006–2011	Yes	717	$1,225.70

[a] For purposes of this article, the 9 chain restaurant companies included in our study were de-identified and assigned a letter, A through I.

[b] *QSR* (10).

[c] Chain restaurants selected as controls were in the following states: Alabama, Arizona, Arkansas, Colorado, Delaware, District of Columbia, Florida, Georgia, Illinois, Indiana, Iowa, Kansas, Kentucky, Louisiana, Maryland, Michigan, Minnesota, Mississippi, Missouri, Montana, North Carolina, North Dakota, Nebraska, New Mexico, Ohio, Oklahoma, Pennsylvania, South Carolina, South Dakota, Tennessee, Texas, Virginia, Utah, West Virginia, Wisconsin, and Wyoming.

We analyzed sections of the menu with the most food items: à la carte adult entrées, adult side dishes, and à la carte children's entrées. Adult entrées and sides (including breakfast items) were identified on the basis of their respective menu section headers, and menu sections that were not consistently included in restaurant nutrition data (desserts, beverages, and condiments) were excluded. To reduce the subjective nature of the classification, we retained items listed on the menu as entrées even if they represented less than what would typically be purchased as an entrée (eg, 1 chicken wing). Children's menu items were identified using the menu section header when available or if the item name contained the words "child" or "kid." Combination items including an entréeand side were excluded, as were items that specified "family" in the item name (eg, popcorn chicken – family size). Nutrient analysis was limited to calories, cholesterol, fiber, saturated fat, and sodium. Although the majority of existing regulations solely require the posting of calorie information, our analysis included additional nutrients in order to broadly characterize the nutritional quality of fast-food menus and to assess key nutrients of concern in the American diet.

Summary nutrition values

Restaurants varied in the number of menu items listed annually. To avoid skewing results toward restaurants with numerous items, we calculated the annual mean for each of the 5 nutrition values by restaurant and used that mean to generate annual means for the case and control restaurants. Three menu years were unavailable (Restaurant H, 2009; Restaurant B, 2005; Restaurant I, 2005) and thus were interpolated linearly from available data for each of the restaurants.

Healthier adult menu items

To determine if restaurant chains altered the proportion of "healthier" menu items, we designated offerings as healthier by using criteria based on Dietary Reference Values (DRV) for a 2000-calorie diet (12), the US Food and Drug Administration standards for packaged food labels (13), and the Dietary Guidelines for Americans (12). There are no federal guidelines for appropriate nutrient levels for adult à la carte restaurant menu items; thus, we selected less than or equal to 25% of the DRV as the criterion for healthier à la carte entrées and less than or equal to 10% of the DRV as the criterion for healthier adult side dishes (Table 2). Some recently published studies used one-third of daily values as a benchmark (9,14) but did not distinguish entrées with and without side dishes. We used one-fourth of daily values because we limited fast-food entrées to those without side dishes (à la carte, not full meals) and because research has shown that most fast-food customers order 2 food items (3). A summary indicator flagged adult menu items that met 4 of 5 nutritional criteria. We originally restricted this indicator to 5 of 5 criteria. However, no menu items satisfied all 5 nutrient criteria, so we relaxed the definition to 4 of 5. Subsequently, we calculated the annual proportion of healthier items for à la carte entrées and sides for each of the 9 restaurants and then aggregated the information for case restaurants and control restaurants.

Table 2. Dietary reference values and healthier[a] limits for fast-food menu items

Nutrition value	Calories (kcal)	Cholesterol (mg)	Fiber (g)	Saturated fat (g)	Sodium[b] (mg)
Adult dietary reference value[c]	2,000	300	25	20	2,300
Healthier limits for adult fast-food menu items[d]					
Entrée, à la carte only, 25% DRV	500	75	6	5	575
Side dish, 10% DRV	200	30	3	2	230
Children's dietary reference value for sedentary/moderately active 8 year old[c]	1,400	210	18	14	1,610
Healthier limits for children's fast-food menu items[e]					
Entrée, à la carte only, 25% DRV	350	50	4	4	400

Abbreviations: DRV, daily reference values.

[a] Healthier is defined as foods with DRV at or below 25% for à la carte entrées and at or below 10% for adult side dishes.

[b] This threshold (2,300 mg) is the tolerable upper limit for persons aged under 40 years and not African-American; the upper limit for most adults is 1,500 mg (12,)

[c] *Dietary Guidelines for Americans, 2010* (12).

[d] This study devised a criteria for healthier fast-food menu items because US Department of Agriculture (USDA) does not provide guidelines for adult meals.

[e] This study devised a criteria for healthier fast-food menu items because USDA only provides guidelines for calories, fat, sodium for children's meals but not each component of those meals.

Children's menus

During the study period, 7 of the 9 restaurants listed children's menus (Table 1). The number of menu items by year and restaurant was small, precluding examination at the restaurant level. Thus, we analyzed children's menu items by pooling all children's items for the case restaurants and pooling all children's items for the control restaurants. We based the criteria for children's menu items on the DRVs for a 1,400-calorie diet, which represents typical calorie needs for sedentary to moderately active children aged 8 years, depending on sex and body size (13); the same criteria have been used by others (6). The US Department of Agriculture, through its school meal programs, has guidelines for calories, fat, and sodium content for children's meals. However, it does not have standards for each component of a menu. Thus, we applied the following criteria: children's à la carte entrées were categorized as healthier, as we defined the term, if they were 25% or less of the DRV. Because few children's items met the "healthier" criteria for fiber or sodium, we used 2 summary indicators, 1 indicator for 3 or more of the 5 nutritional criteria for children's items and another indicator for 4 or more.

Statistical analyses

Descriptive analyses were performed on annual means for all nutritional values at the case and control restaurants, annual proportions of healthier menu items, and the case and control differences between first and last year of observed data. Analyses examined each nutrient for items displayed on 9 restaurant menus over 7 years. We estimated case versus control restaurant differences in trends over time for mean nutritional values and proportion of healthier menu items by using the MIXED procedure in SAS version 9.2 (SAS Institute Inc., Cary, North Carolina). Extreme outliers were controlled for by restricting analyses to sections of the menu with the most food items and excluding entrées with sides and entrées that were combination meals. The data were approximately normally distributed, and means and medians approximated each other; therefore, means were reported. Regression models included all years of data, the number of restaurant entrées or sides dishes, and a random intercept for restaurant. The interaction between case status and an indicator variable for pre–post labeling regulation (before vs after 2008) tested whether the outcome variables differed by case status. This procedure was repeated for adult entrées, adult sides, and children's entrées.

RESULTS

A total of 4,055 menu items met the eligibility criteria from menus of 9 total case and control restaurants during the period from 2005 through 2011. After removing items that were not entrées or sides (n = 168), the total analytic sample was 3,887 items: 2,529 adult à la carte entrées, 1,186 adult sides, and 172 children's à la carte entrées.

Average nutrients in adult menu items

Throughout the study period, across all restaurants, the nutritional profile of fast-food menu items showed high levels of sodium and saturated fat coupled with low levels of fiber (Table 3). When values for an average à la carte entrée and a side dish were combined, they contained 70% and 60% of the daily limit of sodium and saturated fat, respectively. Adult à la carte entrées averaged 450 kcal, 85 mg cholesterol, 3 g fiber, 8 g saturated fat, and 1,100 mg sodium (approximately 2,400 mg sodium per 1,000 calories). Adult sides averaged 300 kcal and 640 mg sodium (or approximately 2,000 mg sodium per 1,000 calories) (Figure). When modeling differences in nutrient averages, regression models found no statistically significant changes over time in nutrient averages and no statistically significant differences between the nutritional averages of case and control restaurants ($P \geq .50$). Case-by-case analyses highlighted the heterogeneity in restaurant trends (Table 3): 3 of 5 labeled restaurants improved their offerings, while 2 of 5 showed no improvement and even launched new options, such as bacon cheeseburgers, that increased average calories by almost 20% and cholesterol by almost 140%.

Adult healthier menu items

During the study period, the overall proportion of healthier adult entrées that met 4 or more of the nutritional guidelines remained less than 25%, with fiber and sodium least likely to meet healthier standards (Table 4). Control restaurants had a lower proportion of healthier items than cases, remaining at approximately 7% over the study period (Figure). In contrast, case restaurants increased the proportion of healthier entrées after labeling regulations: from 13% during years 2005 through 2008, up to 20% by 2011 (regression trend analysis found a mean difference of 5% pre–post 2008 in

Table 3. Distribution of nutrition values for adult à la carte entrées, adult side dishes, and children's à la carte entrées, by fast-food chain restaurant, 2005–2011

Menu Section and Restaurant Case Control Status[a]	2005						2011					
Restaurant	**n**	**Calories[b] (kcal)**	**Cholesterol[b] (mg)**	**Fiber[b] (g)**	**Saturated Fat[b] (g)**	**Sodium[b] (mg)**	**n**	**Calories[b] (kcal)**	**Cholesterol[b] (mg)**	**Fiber[b] (g)**	**Saturated Fat[b] (g)**	**Sodium[b] (mg)**
All chain restaurants	**320**	**446 (191)**	**82 (61)**	**2.8 (3.5)**	**7.5 (4.6)**	**1,098 (541)**	**380**	**452 (192)**	**92 (72)**	**2.1 (1.5)**	**8.4 (5.6)**	**1,087 (446)**
Adult à la carte entrees												
Case chain restaurants	**153**	**419 (192)**	**70 (48)**	**3.4 (5.1)**	**6.5 (3.5)**	**1,061 (580)**	**216**	**422 (186)**	**77 (55)**	**2.1 (1.5)**	**6.6 (4.3)**	**1,008 (414)**
Restaurant A	32	525 (164)	65 (38)	1.3 (0.7)	7.0 (3.4)	713 (236)	42	491 (175)	61 (41)	1.4 (0.7)	7.0 (3.4)	738 (254)
Restaurant B	41	415 (211)	98 (78)	6.5 (9.4)	8.6 (3.0)	1,390 (479)	49	289 (162)	64 (25)	2.9 (1.5)	6.6 (3.0)	1,391 (489)
Restaurant C	12	272 (160)	82 (43)	1.8 (1.9)	5.2 (3.2)	1,236 (713)	16	283 (132)	67 (38)	1.4 (1.6)	4.3 (5.4)	833 (467)
Restaurant D	48	503 (298)	46 (33)	1.8 (1.8)	3.9 (2.6)	763 (453)	71	597 (253)	48 (28)	1.6 (1.5)	3.7 (1.9)	773 (370)
Restaurant E	20	379 (127)	61 (49)	5.6 (11.5)	7.9 (5.1)	1,203 (1,021)	38	451 (207)	147 (145)	3.3 (2.0)	11.2 (7.9)	1,303 (491)
Control chain restaurants	**167**	**474 (190)**	**94 (74)**	**2.2 (2.0)**	**8.5 (5.7)**	**1,135 (503)**	**164**	**481 (199)**	**108 (89)**	**2.1 (1.6)**	**10.3 (6.8)**	**1,166 (477)**
Restaurant F	20	353 (105)	93 (52)	2.3 (2.0)	10.1 (5.0)	1,235 (440)	16	347 (134)	117 (66)	1.3 (1.2)	13.2 (7.4)	1,101 (430)
Restaurant G	64	516 (173)	110 (93)	1.2 (1.4)	10.6 (9.0)	1,198 (539)	51	552 (197)	104 (110)	2.1 (1.2)	9.3 (7.1)	1,322 (455)
Restaurant H	45	513 (269)	121 (126)	2.5 (1.9)	7.9 (5.1)	1,207 (599)	51	511 (226)	126 (118)	2.4 (1.6)	10.0 (6.5)	1,168 (569)
Restaurant I	38	514 (212)	53 (25)	2.8 (2.7)	5.5 (3.6)	898 (432)	46	515 (240)	83 (60)	2.5 (2.3)	8.7 (6.4)	1,073 (455)
Adult side dishes												
All chain restaurants	**149**	**295 (140)**	**12 (17)**	**3.1 (2.3)**	**4.1 (3.2)**	**636 (403)**	**184**	**275 (145)**	**11 (15)**	**2.9 (2.0)**	**3.7 (2.9)**	**640 (402)**
Case chain restaurants	**75**	**288 (140)**	**11 (16)**	**3.5 (2.2)**	**3.4 (2.6)**	**626 (395)**	**99**	**264 (143)**	**10 (14)**	**3.1 (1.8)**	**3.3 (2.7)**	**568 (385)**
Restaurant A	14	462 (204)	16 (29)	3.6 (1.7)	6.0 (5.1)	960 (612)	23	389 (179)	14 (18)	3.2 (1.7)	4.4 (3.2)	953 (517)
Restaurant B	15	136 (68)	3 (5)	2.1 (2.2)	1.3 (1.0)	394 (289)	18	129 (84)	4 (8)	1.9 (1.9)	1.3 (1.6)	343 (283)
Restaurant C	10	185 (90)	9 (12)	2.1 (1.5)	2.2 (1.2)	395 (219)	15	164 (81)	6 (10)	2.0 (1.1)	2.5 (2.2)	359 (204)
Restaurant D	22	351 (159)	14 (16)	4.4 (2.7)	5.0 (3.6)	987 (487)	32	373 (231)	13 (14)	3.2 (1.8)	5.8 (4.7)	777 (521)
Restaurant E	14	307 (182)	12 (18)	5.2 (2.6)	2.6 (2.1)	392 (365)	11	263 (141)	11 (20)	5.1 (2.7)	2.3 (1.7)	409 (400)
Control chain restaurants	**74**	**302 (139)**	**13 (18)**	**2.8 (2.5)**	**4.9 (3.8)**	**647 (412)**	**85**	**287 (146)**	**12 (16)**	**2.7 (2.2)**	**4.2 (3.1)**	**713 (419)**
Restaurant F	14	195 (96)	16 (30)	2.0 (1.6)	2.4 (2.2)	501 (196)	13	200 (98)	18 (31)	2.0 (1.6)	2.6 (2.1)	538 (145)
Restaurant G	36	278 (161)	24 (24)	3.8 (4.5)	5.4 (5.4)	1,073 (571)	44	253 (154)	25 (26)	2.7 (2.4)	4.1 (3.9)	988 (586)
Restaurant H	16	331 (147)	1 (3)	2.6 (2.1)	3.9 (1.8)	525 (320)	18	315 (136)	2 (5)	2.7 (1.6)	3.6 (1.6)	683 (371)
Restaurant I	8	404 (151)	9 (14)	2.6 (1.6)	7.6 (5.9)	490 (560)	10	380 (198)	2 (4)	3.6 (3.0)	6.6 (4.9)	643 (574)
Children's à la carte entrees												
All chain restaurants	**22**	**265 (20)**	**37 (11)**	**1.5 (0.9)**	**5.7 (0.8)**	**796 (91)**	**23**	**313 (62)**	**37 (5)**	**1.4 (0.8)**	**4.6 (0.9)**	**754 (159)**
Case chain restaurants	**12**	**261 (8)**	**31 (10)**	**1.1 (0.5)**	**3.8 (0.3)**	**750 (132)**	**12**	**255 (25)**	**31 (1)**	**1.3 (0.4)**	**3.5 (1.1)**	**638 (95)**
Control chain restaurants	**10**	**491 (152)**	**44 (12)**	**2.0 (1.3)**	**7.6 (1.3)**	**842 (49)**	**11**	**372 (99)**	**43 (10)**	**1.6 (1.3)**	**5.8 (0.8)**	**871 (222)**

[a] For the purposes of this article, the 9 chain restaurant companies included in our study were de-identified and assigned a letter, A through I.

[b] All values are mean (standard deviation) unless otherwise noted.

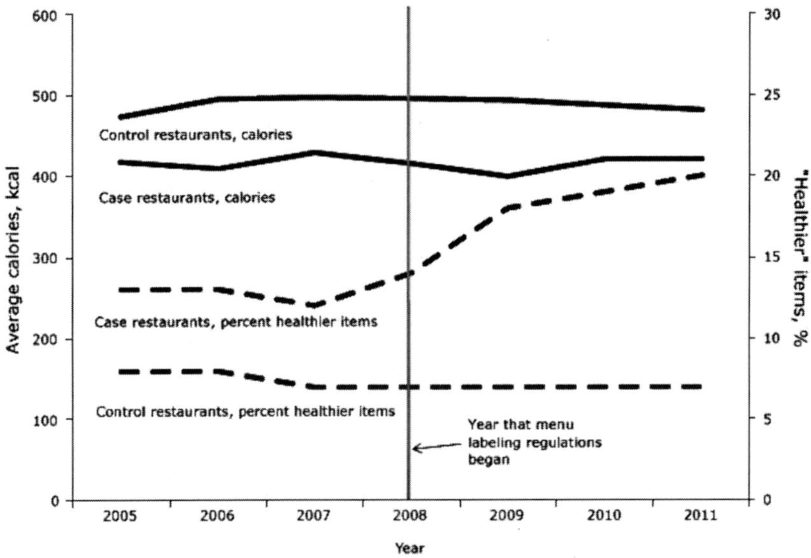

Figure. Average calories for à la carte entrées and percentage of healthier items on fast-food chain restaurant menus, by case and control status and year, 2005–2011. Healthier à la carte entrées are defined as DRV ≤ ...

cases relative to controls: interaction, β = 5.1%; 95% confidence interval [CI], 0.9%–9.3%, P = .02). These improvements were largely attributed to Restaurant A (reduced saturated fat), Restaurant B (offered grilled alternatives to deep-fried entrées, and reduced calories and sodium) and Restaurant D (reduced calories). The prevalence of healthier side dishes was higher among case restaurants than controls (23% vs 15%, respectively) and did not change over time. Similar to entrées, sodium levels in sides were least likely to meet the healthier standard.

Table 4. Percentage of menu items meeting healthier[a] nutritional values for adult à la carte entrees, adult side dishes, and children's à la carte entrees, by fast-food chain restaurant, years 2005 and 2011

Case Control Status[b]	2005						2011					
	Met 4 of 5 criteria (%)	Calories (%)	Cholesterol (%)	Fiber (%)	SaturatedFat (%)	Sodium (%)	Met 4 of 5 criteria (%)	Calories (%)	Cholesterol (%)	Fiber (%)	Saturated fat (%)	Sodium (%)
Adult à la carte entrees	11	66	64	5	41	16	13	67	60	2	41	17
Case chain restaurants	13	70	73	8	50	16	20	70	67	3	53	21
Restaurant A	3	50	59	16	16	3	7	57	76	2	33	7
Restaurant B	12	68	68	2	59	15	27	88	63	0	78	31
Restaurant C	25	92	75	0	83	33	44	94	94	0	88	44
Restaurant D	6	58	77	10	44	10	8	42	46	7	27	8
Restaurant E	20	80	85	10	50	20	13	68	55	5	39	13
	8	62	55	3	32	15	7	64	53	1	28	13
Restaurant F	10	90	60	0	42	35	13	88	75	0	42	38
Restaurant G	6	53	47	8	22	3	4	49	31	0	12	4
Restaurant H	7	56	53	0	29	16	4	65	55	0	33	4
Restaurant I	8	50	61	5	37	8	7	57	50	2	26	7
Adult sides	18	35	87	55	36	22	19	40	89	47	38	18
Case chain restaurants	22	40	89	63	44	20	23	43	89	49	43	23
Restaurant A	0	0	79	64	14	0	4	13	83	61	30	9
Restaurant B	40	87	100	33	87	33	39	78	94	22	78	39
Restaurant C	30	70	100	50	50	20	33	67	100	40	47	27
Restaurant D	14	14	86	86	32	9	13	31	88	50	25	13
Restaurant E	29	29	79	79	36	36	27	27	82	73	36	27
Control chain restaurants	14	31	85	47	28	25	15	37	88	44	33	13
Restaurant F	21	57	86	29	50	7	15	54	85	31	46	0
Restaurant G	14	36	67	53	31	22	18	45	68	36	45	23
Restaurant H	6	19	100	56	19	19	6	28	100	50	22	11
Restaurant I	13	13	88	50	13	50	20	20	100	60	20	20
Children's à la carte entrees[c]	4	71	85	0	34	4	8	69	86	0	38	8
Case chain restaurants	8	92	100	0	58	8	17	92	100	0	67	17
Control chain restaurants	0	50	70	0	10	0	0	45	73	0	9	0

[a] Healthier is defined as foods with daily recommended values at or below 25% for à la carte entrees and at or below 10% for adult side dishes.

[b] For purposes of this article, the 9 chain restaurant included in our study were de-identified and assigned a letter, A through I.

[c] When the definition of healthier children's items was relaxed so that the criteria were 3 or more of the 5 nutrition criteria (rather than 4 or more as shown in the table), prevalence of healthier children's à la carte entrees among all restaurants was 39% (2005), 42% (2011); among case restaurants, 72% (2005), 79% (2011); and among control restaurants, 6% (2005), 6% in (2011).

Children's menu items

Average nutritional values for children's à la carte entrées at case restaurants were 260 kcal and 690 mg sodium and at control restaurants, 430 kcal and 860 mg sodium. Pooled across all restaurants, children's à la carte entrées averaged 37 mg cholesterol, 2 g fiber, and 5 g saturated fat. By using 4 or more of the 5 nutritional criteria to define healthier options, no control restaurants had qualifying items, and only 1 case restaurant had qualifying items (Restaurant E). When healthier options were defined as 3 or more out of the 5 nutritional criteria, the proportion of healthier children's entrées at case restaurants was much higher than at control restaurants. However, no difference was observed in the trend over time: prevalence for meeting 3 of 5 criteria was around 60% for case restaurants and 10% for control restaurants (difference in prevalence, n = 22, $P = .02$; test for interaction in regression trend analysis, $P = .60$). Similar to adult menu items, fiber and sodium in children's menu items were least likely to meet healthier criteria.

DISCUSSION

We found that after the implementation of menu labeling there was a statistically significant increase in the percentage of healthier adult entrées at restaurants in jurisdictions with menu-labeling laws compared with restaurants that were not in jurisdictions subject to labeling. Little improvement, however, was seen among children's entrées during this period, and no significant changes in average nutritional values were seen among adult entrées and sides. Our results suggest menu labeling may provide fast-food restaurants with motivation to introduce healthier menu options; however, greater pressure may be necessary to generate overall average nutritional improvements.

Two recently published articles, by Bruemmer et al (9) and Saelens et al (15), used data from October 2008 through July 2010 to evaluate short-term changes in the nutritional quality of menu items after labeling was enacted in Seattle, King County, Washington. As we found in our study, few substantive improvements were made in fast-food restaurant menus during the study period (9,15). Bruemmer et al's, case-only study design included a reformulation analysis (items that remained on the menu during the study period) that found a statistically significant decrease in fast-food calories, but with small differences; when all fast-food items were analyzed, they found no improvements. In contrast to our study, neither the Bruemmer or Saelens study found improvements in the percentage of fast-food menu options that were healthier (9,15). Differences between those studies and ours included study design, selection of fast-food chains (their studies included more types of fast-food menus: burger, pizza, sandwich, Tex-Mex), sample size, data years, and criteria for healthier items (9,15). We studied a longer time period (2005–2011), including 3 years before New York City implemented its regulations, and we used as our control group chains that had never had to comply with local menu-labeling regulations. To define "healthier" items, other studies used a calorie cut point set to one-third of daily recommended limits (9) or criteria that were specific to a restaurant audit instrument (15), whereas we used a composite index derived from recommendations for calories, cholesterol, fiber, saturated fat, and sodium and set the criteria at one-quarter of daily limits.

To date, menu-labeling regulations have focused primarily on calories. Nutrient profile information reported in this study suggests that policies to address other key nutrients in fast food may have value. At all time periods and at both case and control restaurants, sodium levels remained consistently high, far exceeding healthy limits, and dietary fiber was very low. Diets high in sodium can increase risk of hypertension and cardiovascular disease (16). Dietary fiber has been found to be protective of numerous health conditions (17) and can aid weight reduction by reducing the energy density of foods, while maintaining portion size (18). Fast-food restaurants need to increase fiber in their offerings and be more creative in reformulating their foods to reduce sodium (19).

Study limitations included our reliance on information that fast-food chain restaurant companies posted on their Web pages. However, Web data are quite good representations of actual values: a validation study found only small differences between nutrition information posted on fast-food chain restaurant websites and laboratory analyses (average difference was <10 kcal) (20). A key strength of our study was that we controlled for secular trends by comparing case restaurants with control restaurants (located in jurisdictions outside areas with menu labeling), and we were able to assess the nutritional value of the foods over a long time period (7 years). The restaurant chain companies included in this study have a large presence in the US market, representing more than 23,000 US outlets and reporting $24.7 billion in sales for 2011. However, only 9 chains were eligible because they posted sufficient breadth of historical nutritional information. Therefore, trends observed in this study may not be generalizable to other large chains or different menu types. High heterogeneity within and between restaurants challenged our ability to statistically and substantively assess change. Several steps were taken to reduce high heterogeneity: using menu sections with a large number of menu items that were likely to be similar (example: à la carte entrées), and regression that used restaurant-specific intercepts and controlled for number of items on the menu. Nevertheless, entrée portion sizes varied within and between restaurants, from a single drumstick or fish filet to a large bacon cheeseburger.

In summary, our results in combination with other recently published studies suggest that menu-labeling policies may have already motivated some changes in some chain restaurant menus; however, menus have not markedly improved overall. This study provides a preliminary look at changes in fast-food menus when restaurants first began implementing labeling, but more studies are needed over longer time periods and with larger sample sizes. When menu labeling is required nationally (21), our understanding of industry response to menu labeling will likely become clearer and may be different once all large chains are required to label and customers become accustomed to having nutrition information readily available in chain restaurants. Additional public policies and media advocacy campaigns may be needed to spur broader changes in restaurant offerings so that healthier restaurant choices become the default choice for consumers (22). In addition, policy makers could consider minimum nutritional standards for meals targeted to children, as have been implemented in a few local jurisdictions (23,24). Simultaneous strategies should be considered to encourage chain restaurant companies to significantly improve the nutritional quality of the foods they sell, with portion-size reduction a key focus.

ACKNOWLEDGMENTS

The authors thank Jonathan Cass, Jessica Clark, and Sachin Walawalkar for their assistance with identifying and compiling data for this project and

Charlotte Gerczak for her editorial assistance. This study received funding support from Drexel University and external funding support from the National Institutes of Health, Eunice Kennedy Shriver National Institute of Child Health and Human Development, under grant number U01-HD061978. Mention of trade names, commercial practices, or organizations does not imply endorsement by the authors or the institutions that employ the authors and does not imply endorsement by the funding entities.

Footnotes

The opinions expressed by authors contributing to this journal do not necessarily reflect the opinions of the U.S. Department of Health and Human Services, the Public Health Service, the Centers for Disease Control and Prevention, or the authors' affiliated institutions.

Suggested citation for this article: Namba A, Auchincloss A, Leonberg BL, Wootan MG. Exploratory Analysis of Fast-Food Chain Restaurant Menus Before and After Implementation of Local Calorie-Labeling Policies, 2005–2011. Prev Chronic Dis 2013;10:120224.

REFERENCES

1. Farley TA, Caffarelli A, Bassett MT, Silver L, Frieden TR. New York City's fight over calorie labeling. Health Aff (Millwood)2009;28(6): w1098–109 10.1377/hlthaff.28.6.w1098

2. Bassett MT, Dumanovsky T, Huang C, Silver LD, Young C, Nonas C, et al. Purchasing behavior and calorie information at fast-food chains in New York City, 2007. Am J Public Health2008;98(8): 1457–9 10.2105/AJPH.2008.135020

3. Dumanovsky T, Huang CY, Nonas CA, Matte TD, Bassett MT, Silver LD. Changes in energy content of lunchtime purchases from fast food restaurants after introduction of calorie labeling: cross sectional customer surveys. BMJ 2011;343:d4464 10.1136/bmj.d4464

4. Elbel B, Kersh R, Brescoll VL, Dixon LB. Calorie labeling and food choices: a first look at the effects on low-income people in New

York City. Health Aff (Millwood) 2009;28(6):w1110–21 10.1377/hlthaff.28.6.w1110

5. O'Donnell SI, Hoerr SL, Mendoza JA, Goh E. Nutrient quality of fast food kids meals. Am J Clin Nutr 2008;88(5):1388–95

6. Wellard L, Glasson C, Chapman K. Fries or a fruit bag? Investigating the nutritional composition of fast food children's meals. Appetite 2012;58(1):105–10 10.1016/j.appet.2011.09.024

7. Wootan MG. Children's meals in restaurants: families need more help to make healthy choices. Child Obes 2012;8(1):31–3

8. Rudd, Harris JL, Schwartz MB, Brownell KD. Fast food facts.http://www.fastfoodmarketing.org/researchers.aspx: Yale Rudd Center; 2010. Accessed April 1, 2012.

9. Bruemmer B, Krieger J, Saelens BE, Chan N. Energy, saturated fat, and sodium were lower in entrées at chain restaurants at 18 months compared with 6 months following the implementation of mandatory menu labeling regulation in King County, Washington. J Acad Nutr Diet 2012;112(8):1169–76 10.1016/j.jand.2012.04.019

10. QSR http://www.qsrmagazine.com/reports/top-50-unit-breakdown-sorted-rank Accessed April 1, 2012.

11. Internet Archive Wayback Machine.http://archive.org/web/web.php Accessed April 1, 2012

12. US Department of Agriculture, Dietary Guidelines Committee Dietary guidelines for Americans, 2010. Washington (DC): US Department of Agriculture and US Department of Health and Human Services2011

13. US Food and Drug Administration Guide to nutritional labeling and education act (NLEA) requirements. Washington (DC): Division of Field Investigations, Office of Regional Operations, Office of Regulatory Affairs, US Food & Drug Administration; 1994

14. Wu HW, Sturm R. What's on the menu? A review of the energy and nutritional content of US chain restaurant menus. Public Health Nutr 2013;16(1):87–96

15. Saelens BE, Chan NL, Krieger J, Nelson Y, Boles M, Colburn TA, et al. Nutrition-labeling regulation impacts on restaurant environments. Am J Prev Med 2012;43(5):505–11 10.1016/j.amepre.2012.07.025

16. Appel LJ, Frohlich ED, Hall JE, Pearson TA, Sacco RL, Seals DR, et al. The importance of population-wide sodium reduction as a means to prevent cardiovascular disease and stroke: a call to action from the American Heart Association. Circulation2011; 123(10):1138–43

17. Anderson JW, Baird P, Davis RH Jr, Ferreri S, Knudtson M, Koraym A, et al. Health benefits of dietary fiber. Nutr Rev2009;67(4):188–205

18. Ledikwe JH, Blanck HM, Khan LK, Serdula MK, Seymour JD, Tohill BC, et al. Dietary energy density determined by eight calculation methods in a nationally representative United States population. J Nutr 2005;135(2):273–8

19. Dötsch M, Busch J, Batenburg M, Liem G, Tareilus E, Mueller R, et al. Strategies to reduce sodium consumption: a food industry perspective. Crit Rev Food Sci Nutr 2009;49(10):841–51

20. Urban LE, McCrory MA, Dallal GE, Das SK, Saltzman E, Weber JL, et al. Accuracy of stated energy contents of restaurant foods. JAMA 2011; 306(3):287–93 10.1001/jama.2011.993

21. Stein K. A national approach to restaurant menu labeling: the Patient Protection and Affordable Health Care Act, Section 4205. J Am Diet Assoc 2010;110(9):1280–6, 1288–9

22. Frieden TR. A framework for public health action: the health impact pyramid. Am J Public Health 2010;100(4):590–5]

23. Otten JJ, Hekler EB, Krukowski RA, Buman MP, Saelens BE, Gardner CD, et al. Food marketing to children through toys response of

restaurants to the first US toy ordinance (Santa Clara County CA). Am J Prev Med 2012;42(1):56–60

24. City of San Francisco Health Code. Ordinance 101096 (2010). Setting nutritional standards for restaurant food sold accompanied by toys or other youth focused incentive items.

4

CHAPTER

MEASURING PERCEIVED SERVICE QUALITY OF FAST FOOD RESTAURANT IN ISLAMIC COUNTRY: A CONCEPTUAL FRAMEWORK

Sik Sumaedi [a*], Medi Yarmen [b]

[a,b]*Indonesian Institute of Sciences Puspiptek Area, Building 417, Setu, South Tangerang, 15314, Indonesia*

ABSTRACT

Service quality is a critical success factor of fast food restaurant. Fast food restaurant managers need to measure and improve the service quality of their restaurant continuously. Given this, the knowledge on measurement of service quality of fast food restaurant is needed. Service quality is an abstract and elusive concept. Furthermore, the way of customer on perceiving the quality of service depends on the type of service and the context where the service is provided. Thus, this paper aims to propose a service quality model that is specifically designed for measuring perceived service quality of fast food restaurant in Islamic Country.

Keywords: Service quality, fast food restaurant, Islamic Country, Measurement Model

INTRODUCTION

It is well known that service quality is a critical success factor for a service company [1, 2]. Many empirical studies have shown that service quality positively influences customer satisfaction [3, 4, 5], customer trust [6], customer loyalty [3, 7], and customer perceived value [3]. On the other hand, a service Company that doesn't focus on service quality will be leaved by its customer and even obtains bad reputation [8]. Therefore, a service company should monitor and improve the quality of services the company provides.

Fast food restaurant is a business that can be categorized as a service business [8]. Referring to the explanation on the previous paragraph, a fast food restaurant needs to monitor and improve its service quality in order to win business competition [9, 10]. Furthermore, empirical researches have proved that fast food restaurant service quality positively affects customer satisfaction [11, 12, 13, 14], revisit intention [13,15] and customer's purchasing decision [16].

In order to monitor and improve their fast food restaurants' service quality, the managers of fast food restaurants need to know the way to measure fast food service quality [10, 17]. This is important because service quality is agreed to be an elusive and abstract construct [18, 19]. The failure of a service company in measuring service quality effectively will cause the existence of gap between the perception of the company on service quality and the perception of its customer perceived service quality [8,18]. Furthermore, the gap will make negative customer perceived service quality [8, 18]. Thus, one of the fundamental issues that are important to be discussed is the fast food restaurant service quality measurement [10, 17].

Some researchers have tried to proposed service quality models, such as Servqual model [18], Nordic model [20], Multilevel model [19], and Servperf model [21]. Although the models are widely used by service management researchers, the models are critiqued because the models are generic and need to be adjusted with some services specific factors [22]. For example, Brady and Cronin [19] stated that their multilevel model needs to be adjusted with services specific factors when the model is implemented in a certain type of services. Servqual model is critiqued due to the stability of the service quality dimensions the model proposed [23]. Researchers who re-tested the service quality dimensions of Servqual model found that the number of the dimension of the model can be differed depending on the service industry characteristics [23].

Empirical studies have proved that service quality measurement is affected by service type and the cultural context where the service is provided [24, 25, 26]. Related to this matter, Gayatri et al. [25] and Gayatri and Chew [26] have also found that Islamic values adopted by Muslim customers influence the way of customers in evaluating service quality. This means service quality measurement model of a fast food restaurant is Islamic country should involve service quality dimension that represents service aspect that related with Islamic values adopted by its customers.

In the fast food restaurant context, some researchers have proposed service quality models that are aligned with the characteristics of fast food restaurant and different with the generic service quality model [12, 14]. However, literature service quality that discusses fast food restaurant service quality model that involves service quality dimension that represents service aspect that related with Islamic values is still very limited [9]. Therefore, in order to address the gap in the literature, this paper aims to propose service quality model that is specifically designed for measuring perceived service quality of fast food restaurant in Islamic Country.

LITERATURE REVIEW

Service Quality

Service quality is a topic that is important to be discussed by practitioners and academicians that related with service industry [1, 2]. This is because service quality has strategic roles [1, 2]. In a mature service industry, service quality is an order qualifier that ensures a service company survives

[27]. In other words, service quality is a source of comparative advantage [27]. Meanwhile, in a growth service industry, service quality is an order winner, that is a source of competitive advantage [27].

Service quality has been widely discussed in service management literature [1, 2]. Many researchers have tried to define the construct. Generally, they have agreed that service quality should be studied by using customer perspective [2, 22, 23]. Given this, researchers propose that service quality is customer evaluation on the superiority of service performance provided by a service company [28]. More clearly, Zeithaml [28] explained that service quality is "the consumer's judgment about a [service]'s overall excellence or superiority". Zeithaml [28]'s definition on service quality is the widely quoted definition of service quality [28, 29].

Referring to the previous explanation, it can be stated that service quality is customer evaluation on the superiority of service performance provided by a service company [28]. Thus, fast food restaurant service quality is fast food restaurant customer evaluation on the superiority of the services provided by fast food restaurant [17].

Service Quality Model

To date, there is no agreement on service quality measurement [14, 19]. However, researchers have agreed that service quality is a multidimensional construct [14, 19]. This means service quality consists of more than one dimension. Furthermore, service quality dimension represents service component that is important for customer [2].

Some researchers have proposed generic service quality models. It is believed that the models can be used for measuring service quality of various services [18]. Parasuraman et al. [18] proposed a generic model that is called as Servqual. Servqual reveals that service quality consists of five dimensions, namely tangibles, reliability, assurance, empathy, and responsiveness [18]. Servqual is the service quality model that is widely adopted by service quality researchers [23], including fast food restaurant service quality researchers [10]. However, Servqual has some fundamental problems [23]. First, the model only includes service quality dimensions that can be categorized as functional quality dimension [23]. In other hands, technical quality dimension is excluded [23]. Second, the stability of

Servqual's dimensions is also questioned [23]. This is because the number of Servqual's dimension was changes when researchers retested the model [23].

Another generic service quality model is Nordic model. The model, which is proposed by Gronroos [20], divides service quality into two main dimensions, namely technical quality dimension and functional quality dimension. The technical quality dimension relates with the outcome of services while the functional ones relates with the way on delivering services [20]. Nordic model also involved the important role of corporate image in developing customers' perceived service quality [20].

In addition to Servqual and Nordic model, Brady and Cronin [19] proposed multilevel service quality model. They extended the idea of Gronroos [20] by adding one service quality dimension, i.e. environment quality dimension. The dimension focuses on the servicescape where the services are provided [19]. More specifically, Brady and Cronin [19] revealed that service quality has three dimensions, namely outcome quality, interaction quality, and environment quality. Outcome quality is another name for Gronroos [20]'s technical quality while interaction quality is another name for Gronroos [20]'s functional quality. The idea of Brady and Cronin [19] is also widely adopted by researchers for measuring service quality, including fast food restaurant service quality [14].

Generic service quality models have been proposed by some researchers [18, 19, 20, 21]. Furthermore, the models have also been adopted by other service quality researchers, including fast food restaurant service quality researchers [14]. However, the use of generic service quality model has some problems from both theoretical and practical aspect. Theoretically, researchers have proved that the way of customer on evaluating the quality of service depends on the service contexts, such as service type and customers' culture [24, 25, 26]. In addition, the proposer of the generic quality models also warned users to consider the specific characteristics of the service industry to be measured when they adopts the generic models [19]. Practically, the use of generic service quality model will make the model users will lost specific information of the service they needed [22]. Based on these considerations, fast food restaurant service quality model should be adapted to the context of fast food restaurant.

Fast Food Restaurant Service Quality Model

Defining Fast Food Restaurant

Restaurant is a part of food service industry [10, 17, 30, 31]. Generally, restaurant can be classified into fast food restaurant and full service restaurant [30]. In this paper, fast food restaurant is defined as restaurant that serves fast food, both of which are located in its own building and into a part with other fast food restaurants [31]. Meanwhile, fast food is defined as "food that can be prepared and served very quickly" [10]. Thus, it can be categorized as fast food restaurant, such as McDonalds, KFC, Pizza Hut, Domino Pizza, and Subway [10]. Fast food restaurant business is growing rapidly in this decade [10, 17, 30, 31]. Some multinational companies open their business in other countries aggressively [10]. The companies include McDonalds, KFC, Pizza Hut, Domino Pizza, and Subway [10].

One of the fundamental challenges that must be considered by the companies doing the invasion is the culture of the target market customers [32]. Given this, fast food restaurant that operates in Islamic countries should consider the customer needs on service quality dimension that related with service component representing Islamic values [9, 26].

Fast Food Service Quality Model

Fast food restaurant service quality has been investigated by some researchers. Some researchers adopted Servqual as fast food restaurant service quality measurement model [10, 17, 33, 34, 35]. For example, the study performed by Kaur [10] in Chandigarh, India; Chow and Luk [34] in Toronto, Canada; Oluseye [35] in Nigeria.

Other researchers have tried to develop service quality model that adapted with the specific characteristics of fast food restaurant. For example, Tan et al. [14] proposed Chinese Fast Food Restaurants Service Quality Scale (CFFRSERV). More completely, table 1 shows some service quality models that are specifically designed for measuring service quality in fast food restaurant.

Table 1. Fast Food Service Quality Model

Author (s)	Dimensions
Tan et al. [14]	Assurance and empathy, food cleanliness, responsiveness, reliability, and tangibles
Qin et al. [12]	Recoverability, assurance, tangibles, empathy, responsiveness, and reliability
Wen et al. [11]	Recoverability, assurance, tangibles, empathy, responsiveness, and reliability
Campos and Nobreda [31]	Dimensions were identified to be different in each stage customer must perform to obtain fast food.
Yunus et al. [36]	assurance, tangibles, empathy, responsiveness, reliability, low food prices, short service time and convenient location

Based on the previous explanation, it can be concluded that researchers who discussed fast food restaurant service by involving service quality dimensions that represents the performance of service companies in satisfying Muslim customer needs related to Islamic values is very limited. This is quite surprising since Gayatri et al. [25]'s research shows the service quality measurement model is influenced by Islamic values of Muslim customers. More specifically, Gayatri and Chew [26] revealed that service quality models of Muslim customers should involve general service quality dimensions and Islamic service quality dimensions. Thus, fast food restaurant service quality should consider general service quality dimensions and Islamic service quality dimensions.

Islamic Country

Islam is a comprehensive religion which it not only set the ritual aspect but also regulates the aspects of the daily life of its adherents, including how to dress and how to eat [37]. A Muslim is required to implement Islamic teachings in his daily life even though the country where he/she lives doesn't state Islam as its constitution explicitly. Given this, this research defined Islamic country as Country with Muslim as the majority population and there is no prohibition of the use of Islamic symbol in public area. Based on the definition, Islamic countries includes countries that explicitly stated that Islam as their constitution, such as Saudi Arabia and Brunei Darussalam and countries that do not explicitly state that Islam as their constitution but its majority population is Muslim and there is no prohibition of the use of Islamic symbol in public area, such as Indonesia and Malaysia.

FAST FOOD RESTAURANT SERVICE QUALITY MODEL IN ISLAMIC COUNTRY: A CONCEPTUAL FRAMEWORK

Based on Brady and Cronin [19], service quality model should involve three aspects, namely outcome, environment, and interaction. Outcome relates to the result of the service provided; interaction refers to the service delivery; and environment relates to the servicescape where the service provided [19]. The view of Brady and Cronin [19] was supported by some restaurant service quality researchers, including fast food restaurant service quality researchers [14]. Hence, fast food restaurant service quality dimensions should represent these three aspects. According to Gayatri and Chew [26], service quality measurement is influenced by Islamic values had by the customers. Furthermore, they revealed that Muslim customers view service quality from two aspects, namely general service quality dimension and Islamic service quality dimension [26]. Therefore, fast food restaurant service quality model in Islamic country should represent general service quality aspect and Islamic service quality aspect. Based on the explanation, this paper proposed fast food restaurant service quality model in Islamic country as shown in figure 1.

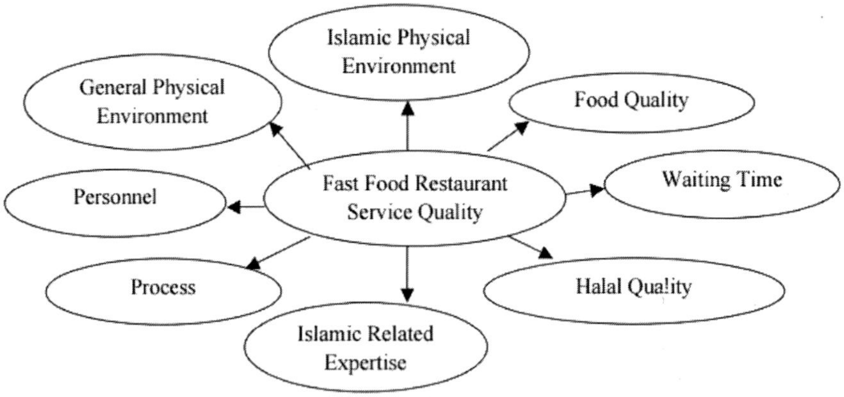

Figure 1. Fast Food Service Quality Model for Islamic Country

Table 2. The Operational Definition of Service Quality Dimension

Brady & Cronin [19]'s Category	Gayatri & Chew [26] 's Category	Service Quality Dimension	Operational Definition	Related Literature
Environment quality	General Service quality	Physical Environment	Physical facilities and equipment used to provide the services	[9], [10], [14], [18]
	Islamic Service quality	Islamic Physical Environment	Physical facilities and equipment that specifically designed for fulfilling Customers' Islamic value related needs	[25], [26]
Outcome quality	General Service quality	Food quality	The taste and quality of the food provided	[12], [14]
		Waiting time	The time sacrificed by customer for waiting the food being served from he/she arrives	[19]
	Islamic Service quality	Halal Quality	The halal status of the food provided	[25], [26]
Interaction quality	General Service quality	Personnel	The attitude, behavior, and appearance of personnel	[9], [10], [14], [18], [19]
		Process	The process arranged for delivering the services and interacting with customers	[9], [10], [14], [18], [19]
	Islamic Service quality	Islamic Related Expertise	Willingness to help and ability to satisfy customer inquiries regarding their Islamic value	[25], [26]

The operational definition of each proposed service quality dimension can be seen in table 2. Based on table 2, it can be seen that the proposed fast food restaurant service quality model in Islamic country consists of eight dimensions. Based on the classification of Gayatri and Chew [26], the model involves five general service quality dimensions and three Islamic service quality dimensions. Based on Brady and Cronin [19]'s categorization, the proposed service quality model has three outcome quality dimensions, which consist of two general service quality dimensions (waiting time, food quality) and one Islamic service quality dimension (Halal quality). The proposed model also has two environment quality dimensions, which includes one general service quality dimension (general physical environment) and one Islamic service quality dimensions (Islamic physical environment).

Furthermore, the proposed model involves three interaction quality dimensions, which includes two general service quality dimensions (personnel, process) and one Islamic service quality dimension (Islamic related expertise).

FUTURE RESEARCH METHODOLOGY

The proposed fast food restaurant service quality model in Islamic country was developed based on literature review. Therefore, future research will identify the indicators of the proposed model dimension and test the proposed fast food restaurant service quality model. Based on Churchill [38], we recommend some steps that can be performed in future research.

The first step of the research aims to identify the appropriate indicators for measuring the proposed service quality model dimensions. To achieve the objective of the step, it is needed to perform qualitative research by using in depth interview with the customers of fast food restaurant service quality in Islamic country. In addition, it is needed to perform Focus Group Discussion involving the customers in order to verify and clarify the indicators obtained from the in depth interview.

The second step is the content validity testing of the indicators obtained in the first step. This step aims to check whether the indicators used in the research are appropriate to measure the proposed service quality dimension's concept. Given this, each indicator will be reviewed by two service management expert. The indicator which according to the experts is not in accordance with the proposed concept is eliminated. The result of the second step is a set of indicators that can be used to develop an instrument for measuring the quality of services based on the proposed fast food restaurant service quality model.

The third step is the collection of empirical data. This will be done through survey using an instrument of questionnaire that has been developed in the previous step. The respondents of survey will be 200 Muslims customers of fast food restaurant. The number of the respondent is determined based on the statistical analysis requirements that will be used in the next step, namely Structural Equation Modeling (SEM) [39].

The fourth step is the model's goodness of fit, construct validity, criterion-related validity, and reliability testing. The goodness of fit of the model and construct validity testing will be performed by using SEM. Construct validity investigated will consist of discriminant validity and convergent validity [39, 40]. Criterion-related validity testing will be performed by analyzing the correlations between the proposed dimensions of service quality model with the constructs theoretically relate to service quality, namely loyalty and satisfaction [39, 40]. Reliability testing will be performed by analyzing Cronbach alpha coefficient [39, 40].

CONCLUSION

Service quality is a key success factor for fast food restaurant. Given this, the knowledge related to the conceptualization and measurement of fast food restaurant service quality becomes important. This paper aims to propose a service quality model that can be used to measure fast food restaurant service quality in the Islamic country. The proposed fast food restaurant service quality model consists of eight dimensions. More specifically, the proposed service quality model has three outcome quality dimensions, which consist of two general service quality dimensions (waiting time, food quality) and one Islamic service quality dimension (Halal food). The proposed model also has two environment quality dimensions, which includes one general service quality dimension (general physical environment) and one Islamic service quality dimensions (Islamic physical environment). Furthermore, the proposed model involves three interaction quality dimensions, which includes two general service quality dimensions (personnel, process) and one Islamic service quality dimension (Islamic related expertise).

REFERENCES

1. Yusoff, W.Z.W., Ismail, M. and Ali, A.S. Understanding the Services Provider Perspective towards Better Service Quality in Local Authorities. Journal of Facilities Management 2010; 8 (3): 226-230.

2. Clemes, M. D., Gan, C., Kao, T, H., and Choong M. An Empirical Analysis of Customer Satisfaction in International Air Travel. Innovative Marketing 2008; 4: 50-62.

3. Lai, W. and Chen, C. Behavioral Intention of Public Transit Passenger – the Role of Service Quality, Perceived Value, Satisfaction and Involvement. Transport Policy 2011; 18: 318- 325.

4. Bakti, I.G.M.Y and Sumaedi,S. An Analysis of Library Customer Loyalty: The Role of Service Quality and Customer Satisfaction, A Case Study in Indonesia. Library Management 2013; 34 (6/7): 397 – 414.

5. Olorunnivo, F., Hsu, M. K., and Udo, G. J. Service Quality, Customer Satisfaction, and Behavioral Intention in the Service Factory. Journal of Services Marketing 2006; 20 (1): 59- 72.

6. Moliner, M.A. Loyalty, Perceived Value and Relationship Quality in Healthcare Services. Journal of Service Management 2009; 20 (1): 76-97.

7. Wen, C.H., Lan, L.W., and Cheng, H.L. Structural Equation Modeling to Determine Passenger Loyalty toward Intercity Bus Services. Journal of the Transportation Research Board 2005; 1927: 249 – 255.

8. Zeithaml, V., Bitner, M., and Gremler, D. Services marketing: Integrating Customer Focus across the Firm. 5th ed. Boston: McGraw-Hill; 2009

9. Ibrahim, S. and Othman, M. Developing and Validating Halal Service Quality Instrument for Malaysian Food Service Establishments: A Conceptual Paper. Procedia – Social and Behavioral Sciences 2014; 130: 400-408

10. Kaur, M. Service Quality in International Fast Food Chains in Chandigarh – Perception of Customers. International Journal of Applied Research and Studies 2013; 2 (7): 1-13

11. Wen, C., Qin, H., Prybutok, V.R., and Blankson, C. The Role of National Culture on Relationship between Customers' Perception of Quality, Values, Satisfaction, and Behavioral Intentions. Faculty Research & Creative Activity 2012; paper 9. http//thekeep.eiu.edu/business_fac/9 Sik Sumaedi and Medi Yarmen

12. Qin, H., Prybutok, V.R., and Zhao, Q. Perceived Service Quality in Fast Food Restaurants: Empirical Evidence from China. International Journal of Quality and Reliability Management 2010; 27(4): 424-437

13. Law, A.K.Y., Hui, Y.V., and Zhao,X. Modeling Repurchase Frequency and Customer Satisfaction for Fast Food Outlets. International Journal of Quality and Reliability Management 2004; 21(5): 545-563

14. Tan, Q., Oriade, A., and Fallon, P. Service Quality and Customer Satisfaction in Chinese Fast Food Sector: A Proposal for CFFRSERV. AHTR 2014; 2(1): 30-53

15. Nezakati, H., Kuan, Y.L. and Asgari, O. Factors Influencing Customer Loyalty towards Fast Food Restaurants. International Conference on Sociality and Economics Development 2011;10: 12-16

16. Ying San, Lim, Chen Ho, Tan, and Nya Ling, Tan. The Trend of Healthy Fast Food: How Malaysian' Generation Y Makes the Purchase Decision? Australian Journal of Basic and Applied Sciences 2013; 7(7): 631-638

17. Salami, C.G.E. and Ajobo, R.T. Consumer Perceptions about Fast Food Restaurants in Asaba. Global Journal of Management and Business Research 2012; 12(1): 75-81

18. Parasuraman, A., Zeithaml, V. A., and Berry, L.L. SERVQUAL: A Multi-Item Scale for Measuring Consumer Perceptions of the Service Quality. Journal of Retailing 1988; 64(1): 12- 40.

19. Brady, M.K. and Cronin, J.J. Some New Thoughts on Conceptualizing Perceived Service Quality: A Hierarchical Approach. The Journal of Marketing 2001; 65(3): 34 – 49.

20. Gronroos, C. A Service Quality Model and its Marketing Implications. European Journal of Marketing 1984; 18(4): 36 – 44

21. Cronin, J and Taylor, S A. Measuring Service Quality: A Reexamination and Extension," Journal of Marketing 1992; 56: 55-67.

22. LeBlanc, G. and Nguyen, N. Searching for excellence in business education: an exploratory study of customer impressions of service quality.International Journal of Educational Management 1997; 11(2): 72 – 79

23. Nagata, H., Satoh, Y., Gerrard, S., and Kytömäki, P. The dimensions that construct the evaluation of service quality in academic libraries. Performance Measurement and Metrics 2011; 5 (2): 53 – 65

24. Micuda, D. and Cruceru, G. Testing SERVQUAL dimensions on the Romanian banking sector. Proceedings of the International Conference on Risk Management, Assessment and Mitigation 2010; 329 – 334 130 S

25. Gayatri , G. , Hume, M., and Mort, G.S. The role of Islamic Culture in service Quality research. Asian Journal on Quality 2011:,12 (1): 35 - 53

26. Gayatri, G. and Chew , J. How do Muslim consumers perceive service quality ? Asia pacific Journal of Marketing and Logistics 2013; 25 (3): 472-490

27. Prajogo, D.I., and Sohal, A.S. The relationship between TQM practices and innovation performance: a literature review and research framework. Technovation 2001; 21(9): 539– 558.

28. Zeithaml, V.A. Consumer Perceptions of Price, Quality and Value: A Means and Model and Synthesis of Evidence. Journal of Marketing 1988; 52: 2-22.

29. Park, J.W., Robertson, R., and Wu, C.L. Investigating the Effects of Airline Service Quality on Airline Image and Passengers' Future Behavioural Intentions: Findings from australian international air passengers. The Journal of Tourism Studies 2005; 16(1): 2 –11

30. Ahmad, F., Ghazali, H., and Othman, M. Consumers Preference between Fast Food Restaurant and Casual Dining Restaurant: A Conceptual Paper. 3rd International Conference On Management (3rd ICM 2013) Proceeding 2013; 315-325

31. Campos, D.F.and Nobrega, K.C. Importance and the Zone of Tolerance of Customer Expectations of Fast Food Services. Journal of Operations and Supply Chain Management 2009; 2(2): 56 – 71

32. Mohammad, T. Attributes Influencing Preference of University Students on Fast Food Outlets. Tourism Educators Association of Malaysia 2004; 1(1): 77-88

33. Tat, H.H., Sook-Min, S., Ai-Chin, T., Rasli,A., and Abd Hamid,A.B. Consumers' Purchase Intentions in Fast Food Restaurants: An Empirical Study on Undergraduate Students. International Journal of Business and Social Science 2011; 2(5): 214-221

34. Chow, C.C. and Luk, P. A strategic service quality approach using analytic hierarchy process. Managing Service Quality 2005; 15(3): 278-289

35. Oluseye, O.O. A comparative study of male and female perceptions of service quality in fast food industry. Economic, social, political and cultural problems of the future society 2009; 10: 60-69

36. Yunus, N.K.Y., Razak, M.Z., and Ilias, A. Expectation towards McDonald's Malaysia: A Study on Service Quality. International

Journal of Independent Research and Studies – IJIRS 2013; 2(3): 119-129

37. Ash-Shiddieqy, T.M. The Introduction of Islamic Law (in Bahasa Indonesia). 6th Ed. Jakarta: Penerbit Bulan Bintang; 1980.

38. Churchill, G.A. Jr. A paradigm for developing better measures of marketing constructs. Journal of Marketing Research 1979; 16(1): 64-73.

39. Hair, J.F. Black, W. C., Babin, B. J., and Anderson, R. E. Multivariate data analysis, 7th ed., New Jersey: Prentice-Hall; 2010.

40. Sekaran, U and Bougie,R. Research Methods for Business: A Skill Building Approach. UK: John Wiley and Sons Ltd.; 2011.

5

CHAPTER

FOODSERVICE QUALITY: IDENTIFYING PERCEPTION INDICATORS OF FOODSERVICE QUALITY FOR HOSPITALITY STUDENTS

Wen-Hwa Ko, Li-Jung Su

Department of Restaurant, Hotel and Institutional Management, Fu Jen University, New Taipei City, Taiwan.

ABSTRACT

This study uses an expert perspective to identify indicators of foodservice quality as perceived by hospitality students. A 3-round modified Delphi method and an expert panel comprising culinary arts instructors, master chefs with >10 years of experience, and senior foodservice industry executive managers were used to define and identify these indicators. The developed foodservice quality framework includes 2 categories, the product and customer, which comprise 57 indicators in 6 dimensions. The results of this study provide a clear direction for further studies and applications for hospitality professionals, educators, and students.

Keywords: Foodservice Quality; Hospitality Student; Modified Delphi Method

1. INTRODUCTION

Foodservice quality is a critical factor influencing the hospitality and dining experience [1] and is a concept that hospitality students must understand and master before beginning their careers in the foodservice industry. Each aspect of foodservice quality must be addressed to satisfy the expectations and needs of the consumer [2]. Furthermore, ensuring the quality of both the food and service is critical for the continuing success of restaurants and foodservice operations [3]. The researchers used a modified Delphi approach, including three anonymous expert evaluation rounds and a survey, to identify the primary indicators of hospitality student foodquality perceptions. The researchers used qualitative statistical and content analyses to validate these perception indicators.

2. LITERATURE REVIEW

The food and beverage (F & B) sector encompasses numerous dimensions, including historical and cultural factors, product development, F & B service, design, product evaluation, physiology, nutrition, psychology, markets, health, environment and space, interpersonal relations, product characteristics, and comprehensive management systems [4]. Foodservice quality requirements that are essential for continued innovation as well as for improved reliability and consistency include consumer-specific preferences, such as nutrition, presentation, and taste, and general performance requirements, such as uniform quality, on-time delivery, product differentiation, and research-and-development (R & D)-based improvements [5,6]. Furthermore, food hygiene is an increasingly important factor that affects consumer trust in the overall food quality [7]. Finally, food safety is of paramount importance to the consumer, food industry, and economy [8].

Therefore, learning about and implementing proper foodservice operations are critical to hospitality education. Hospitality education prepares students for future management positions using curricula that teach hospitality-related technical skills and provide a diverse grounding in business and the social sciences. Coursework is often designed in collaboration with the industry, and practical experience internships are a critical part of the hospitality-education experience [9]. The hospitality industry remains a highly labor-intensive sector of the economy; human resources and human-related factors easily contribute 90% or more to the success or failure of hospitality industry operations [10]. Rapid growth in the hospitality sector and foodservice industry in Taiwan has led to an increased awareness of the importance of foodservice quality. Comprehensive quantitative standards and total-quality-assurance systems implemented by many foodservice operators have achieved significant gains in foodservice quality. Consequently, Taiwan's higher education institutions have opened and expanded related academic programs in hospitality and foodservice management as well as marketing, tourism, and restaurant marketing. Professional foodservice and culinary education is an important and growing segment of higher education in Taiwan. Most hospitality students have industrial internship experience before graduation [11].

3. RESEARCH METHODOLOGY

3.1. Study Design and Questionnaire

The Delphi method is a structured approach toward collecting and distilling expert knowledge using a series of questionnaires interspersed with controlled-opinion feedback [12]. To identify and validate the perception indicators, we used a modified Delphi method rather than a traditional Delphi method because the traditional method uses only open questions. The conceptual framework and Delphi survey format were developed for this study under the guidance of the domestic and international literature regarding the hospitality industry foodservice quality, relevant theories, and prior studies on factors affecting the perceptions of foodservice quality held by hospitality and culinary students. The survey included several open questions designed to collect information on new potential foodservice quality indicators. Because foodservice quality addresses interactions with consumers in addition to the physical products, this study considered foodservice quality in terms of six distinct dimensions from two different perspectives: the product and the customer. The dimensions under the product perspective included food character, culinary arts, hygiene, and safety. The dimensions under the customer perspective included environment, ambience, marketing, promotion, and service quality.

The researchers reviewed the literature and generated a preliminary foodservice quality survey instrument based on the relevant findings. Experts were subsequently invited to participate in a series of questionnaire surveys. Invitations were delivered by mail in an opaque envelope, and the recipients were asked to return the enclosed survey questionnaire in the enclosed return envelope within two weeks. During the first round, the survey participants ranked their opinion of each item on a five-point Likert-type scale ranging from "strongly disagree" to "strongly agree". Participants were also invited to provide additional comments to their responses or additional remarks. A report on first-round analysis results was sent to the participants with an invitation to participate in rounds 2 and 3.

3.2. The Expert Panel

Delbecq, Van de Van, & Gustafson [12] stated that 10 to 15 participants on an expert panel consisting of a homogeneous group should represent an adequate sample. This study enrolled 20 participants in three categories: culinary course instructors, master chefs with >10 years of experience, and senior foodservice industry executive managers. All participants had the practical teaching and industry experiences necessary for a substantive understanding of foodservice quality perceptions prevalent in the foodservice industry and among students.

3.3. Standards for Item Elimination

At the end of the first round, items with average scores below 4.00 and quartile deviations above 1.00 were eliminated. After the second and third rounds, those with average scores below 4.00 were eliminated, and amendments or modifications were performed on the items to conform with participant input [13].

3.4. Standard for Consensus

According to Faherty [13] and Holden and Wedman [14], quartile deviations below 0.60 represent a substantial degree of expert consistence, quartile deviations between 0.60 to 1.0 represent moderate consistency, and quartile deviations >1.0 represent a lack of consensus. Holden and Wedman [14] went further, stating that item stability is achieved and further modification may be discontinued when >85% reach a moderate or better consensus.

4. RESULTS

4.1. First Round

The first-round Delphi results were analyzed to obtain the average, mode, and quartile deviation scores. Items with average scores below 4.00 and quartile deviations equal to or greater than 1 were eliminated. Although the average marketing and promotion dimension scores fell below 4.00 (3.85), this dimension was retained in the second round after discussions with experts and the performance of the subsequent amendments. All other categories and dimensions were preserved without amendment. Some indicators were modified, and new indicators were added for the second-round Delphi process based on participant input. The second Delphi-round instrument included 2 categories, 6 dimensions, and 79 indicators.

4.2. Second Round

The second-round instrument was a semi-open-ended questionnaire developed by eliminating the items noted above and amending the remaining items based on participant suggestions. The results of the second round achieved higher average scores and lower quartile deviations than the first, indicating that the second-round modified-Delphi method was more appropriate than the first; participant opinions trended toward uniformity. Based on these results, the researchers made further modifications based on participant suggestions and integrated substantially similar items into the third-round modified Delphi instrument.

The scores for the second modified Delphi instrument round were analyzed to obtain the average, mode, and quartile deviations, and indicators with average scores below 4.00 were eliminated. Although the marketing and promotion dimension achieved average scores below 4.00, its importance and expert consistency were higher than the levels achieved in the first round, with 7 indicators in this domain earning average scores ≥4.00. Therefore, this dimension was retained and led to 7 indicators being included in the third round. The dimensions and categories of the remaining indicators were not affected, although amendments to some indicators were performed based on the suggestions of the participants. Consequently, the third modified Delphi instrument had 2 categories, 6 dimensions, and 62 indicators.

4.3. Third Round

The third-round modification of the instrument added several new indicators, while dropping or merging several second-round indicators. A comparison of the appropriateness of each category appears in Table 1, indicating that the two categories, the product and customereach earned average scores above 4.80 and quartile deviations of 0. This result evinced a high degree of expert agreement regarding the importance of these two foodservice quality categories.

The three indicators in the product category earned average scores above 4.00 and quartile deviations below 0.6. Thirty indicators achieved high consistency; 16 earned scores of 5 (highly significant), and 14 earned scores of 4 (important). Indicators including "wellmatched condiments" and "possessing professional culinary arts qualifications" both earned average scores below 4.0, justifying their elimination. Although "product calorie information display" failed to achieve an average score of 4.0, the item was amended based on expert suggestions and retained as an indicator under a new item name: "nutritional information display". The third-round results and expert-recommended changes to some of the secondary indicators provided the basis for the final set of foodservice quality perception indicators for the hospitality sector.

Two of the three indicators in the customer category earned average scores above 4.00, while the marketing and promotion dimension earned an average of 3.95. The quartile deviations for the three primary indicators were all below 0.6. Thirty-two indicators achieved high consistency: 17 earned scores of 5 (highly significant), and 15 earned scores of 4 (important). Although the thirdround score for the marketing and promotion dimension fell short of 4.00 by averaging 3.95 with a quartile deviation of 0, the score was higher than the first (3.85) and second (3.90) rounds, demonstrating increased importance and momentum toward expert consensus. Five of this dimension's indicators attained average scores of ≥4.00. Marketing and promotion were therefore retained with the other five dimensions.

Table 1. Statistical results for the two categories.

Categories	Round 2			Round 3		
	Average	Mode	Quartile Deviation (Q.D.)	Average	Mode	Quartile Deviation (Q.D.)
Product orientation	4.75	5	0.125	4.85	5	0
Customer orientation	4.85	5	0	4.80	5	0

Table 2. Statistical results for the six dimensions.

	2nd round			3rd round		
Dimensions	Average	Mode	Quartile Deviation (Q.D.)	Average	Mode	Quartile Deviation (Q.D.)
Product character	4.68	5	0.5	4.90	5	0
Culinary arts	4.30	4	0.5	4.30	4	0.5
Hygiene and safety	4.95	5	0	4.90	5	0
Environment atmosphere	4.10	4	0	4.25	4	0.125
Marketing and promotion	3.90	4	0	3.95	4	0
Service quality	5.00	5	0	5.00	5	0

Table 2 compares the results for the primary indicators across the second and third rounds, with the final list including 57 indicators in 6 dimensions in 2 categories.

Table 3. Final list of the perception indicators of foodservice quality.

I. Product category

1. Product character
(1) appearance and color
(2) texture
(3) aroma
(4) flavor
(5) freshness
(6) source
(7) plating and matching
(8) temperature
(9) nutritional information display

2. Culinary arts
(1) nutritional knowledge
(2) menu planning knowledge and skills
(3) culinary arts creativity
(4) menu innovation skills
(5) ingredient and culinary method pairing skills

3. Hygiene and safety

(1) product cleanliness and hygiene
(2) product-source inspection certificates
(3) periodic product inspections
(4) kitchen cleanliness and hygiene
(5) kitchen operations flow planning
(6) pantry storage administration
(7) employee personal-health and hygiene management
(8) waste handling and disposal methods
(9) understanding hygiene regulations
(10) packaging process hygiene and safety
(11) kitchen facility routine disinfections
(12) drinking water and ice hygiene safety inspections

II. Customer category

1. Environment atmosphere
(1) dining area air quality
(2) dining area lighting and brightness
(3) dining area cleanliness and hygiene
(4) tablecloth and utensil cleanliness
(5) tablecloth arrangement and design

Continued

(6) staff customer interaction vocal volume
(7) dining area space planning
(8) dining area seating comfort
(9) restroom equipment and cleanliness
(10) dining area overall ambience management
(11) background music selection appropriateness
(12) fire equipment periodic inspections
(13) escape route maps and escape indicators clearly marked
2. Marketing and promotion
(1) price fairness
(2) product value exceeds cost
(3) advertising imagery conforms to actual performance
(4) serving size appropriateness
(5) corporate social responsibility and brand image
3. Service quality
(1) staff apparel and appearance/etiquette
(2) staff service attitudes
(3) communication and responsiveness skills
(4) menu explanation abilities
(5) adequate professional foodservice knowledge
(6) linguistic skills
(7) respect for customer privacy
(8) respect for customer opinions/demands
(9) service timeliness
(10) ability to handle challenges/unexpected situations
(11) service accuracy
(12) response and handling of customer complaints
(13) staff marketing and promotional talents

5. DISCUSSION AND RECOMMENDATIONS

Expert opinions indicate that the importance of the "customer orientation" dimension exceeds that of the "product orientation" dimension. Chen and Huang [15] demonstrated that when selecting dining venues, consumers increasingly emphasize food quality and price, dining setting and ambience, and service quality, reflecting their greater emphasis on food hygiene and safety, interpersonal interactions, and the importance of personal preferences and values. Interior design and aesthetics can accentuate a restaurant's unique dining ambiance and style. The design of building exteriors and interiors [16,17], the creation of a distinctive ambience [18,19], and the development of visually striking menu presentations [20] all play positive roles. Therefore, restaurant operators must search for new experiential elements to enhance customer excitement and stimulate a customer's willingness to patronize and re-patronize [17,21].

The overlap between the literature and the expert cohort findings demonstrates the importance of foodservice quality dimensions throughout society. The firstand second-round as well as the secondand third-round modified Delphi questionnaire stability exhibited high degrees of stability among the participant opinions. This finding validates and completes this study's modified Delphi instrument. The final version included 2 categories (product and customer), 6 dimensions (product character, culinary arts, hygiene and safety, environment and ambience, marketing and promotion, and service quality), and 57 perception indicators of food quality (Table 3).

The final version may be used to enhance the teaching of restaurant management and food preparation principles to improve students' food-related knowledge and experience. The final version may also be used to develop practical kitchen courses that foster a greater appreciation of foodservice quality. As consumer foodservice quality expectations continue to rise, a restaurant's success requires not only the provision of excellent food but also achievements in other dimensions. Based on study findings, we recommend that the hospitality industry focuses greater attention on menu item prices, hygiene, environment, and service conditions to enhance the foodservice quality experience of customers. Additionally, internships and on-the-job training programs should reduce repetitive task assignments and expand management-related training to expose students to all facets of the restaurant industry, enhance a student's willingness to enter the field after graduation, and improve overall foodservice quality perception performance. The Delphi survey in this study used input from experts to construct foodservice quality perception indicators. Future studies should survey the opinion of restaurant employees, hospitality teachers, and students to confirm the validity of the identified indicators.

REFERENCES

1. J. M. Sulek and R. L. Hensley, "The Relative Importance of Food, Atmosphere, and Fairness of Wait," Cornell Hotel and Restaurant Administration Quarterly, Vol. 45, No. 3, 2004, pp. 235-247.

2. Y. Namkung and S. Jang, "Does Food Quality Really Matter in Restaurant: Its Impact of Customer Satisfaction and Behavioral Intentions?" Journal of Hospitality and Tourism Research, Vol. 31, No. 3, 2007, pp. 387-410.

3. J. R. Olsen, H. Harmsen and A. Friis, "Linking Quality Goals and Product Development Competences," Food Quality and Preference, Vol. 19, No. 1, 2008, pp. 33-42.

4. H. L. Meiselman, "Dimensions of the Meal," Journal of Foodservice, Vol. 19, No. 1, 2008, pp. 13-21.

5. S. Sanderson and M. Uzumeri, "Managing Product Families: The Case of the Sony Walkman," Research Policy, Vol. 24, No. 6, 1995, pp. 762-782.

6. C. Y. Baldwin and K. B. Clark, "Managing in an Age of Modularity," Harvard Business Review, Vol. 75, No. 5, 1997, pp. 84-93.

7. D. Aggelogiannopoulos, E. H. Drosinos and P. Athanasopoulos, "Implementation of a Quality Management System (QMS) According to the ISO 9000 Family in a Greek Small-Sized Winery: A Case Study," Food Control, Vol. 18, No. 9, 2007, pp. 1077-1085.

8. M. Jevsnik, V. Hlebec and P. Raspor, "Consumers' Awareness of Food Safety from Shopping to Eating," Food Control, Vol. 17, 2008, pp. 737-745.

9. C. W. Barrow and R. H. Bosselman, "Hospitality Management Education," The Haworth Hospitality Press, 1999.

10. C. C. Teng, "The Effects of Personality Traits and Attitudes on Student Uptake in Hospitality Employment Student Uptake in

Hospitality Employment," International Journal of Hospitality Management, Vol. 27, No. 1, 2008, pp. 76-86.

11. J. S. Horng, "Curriculum Analysis of Foods and Beverage management in Technological and Vocational Education," Asia Pacific Journal of Tourism Research, Vol. 9, No. 2, 2004, pp. 107-119.

12. A. L. Delbecq, A. H. Van de Ven and D. H. Gustafson, "Group Techniques for Program Planning: A Guide to Nominal Group and Delphi Processes," Scott, Foresman, Glenview, 1975.

13. V. Fahety, "Continuing Social Work Education: Result of Delphi Survey," Journal of Education for Social Work, Vol. 15, No. 1, 1979, pp. 12-19.

14. M. C. Holden and J. F. Wedman, "Future Issues of Computer-Mediated Communication: The Results of a Delphi Study," Educational Technology, Research and Development, Vol. 41, No. 4, 1993, pp. 5-24.

15. K. H. Chen and D. H. Huang, "Devising Evaluation Indicators for Taiwanese Cultural Cuisine Restaurant Rating," Journal of Hospitality and Home Economics, Vol. 7, No. 3, 2010, pp. 235-259.

16. D. J. MacLaurin and T. L. MacLaurin, "Customer Perceptions of Singapore's Theme Restaurants," Cornell Hotel and Restaurant Administration Quarterly, Vol. 41, No. 3, 2000, pp. 75-85.

17. K. Ryu, H. Han and T. H. Kim, "The Relationships among Overall Quick-Casual Restaurant Image, Perceived Value, Customer Satisfaction, and Behavioral Intentions," International Journal of Hospitality Management, Vol. 27, 2008, pp. 459-469.

18. R. Weiss, A. H. Feinstein and M. Dalbor, "Customer Satisfaction of Theme Restaurant Attributes and Their Influence on Return Intent," Journal of Foodservice Business Research, Vol. 7, No. 1, 2004, pp. 23-41.

19. Y. Liu and S. Jang, "Perceptions of Chinese Restaurants in the US: What Affects Customer Satisfaction and Behavioral Intentions?"

International Journal of Hospitality Management, Vol. 28, No. 3, 2009, pp. 338-348.

20. J. Kivela, R. Inbakaran and R. Reece, "Consumer Research in the Restaurant Environment, Part 1: A Conceptual Model of Dining Satisfaction and Return Patronage," International Journal of Contemporary Hospitality Management, Vol. 11, No. 5, 1999, pp. 205-222.

21. Y. Liu and S. Jang, "The Effects of Dining Atmospherics: An Extended Mehrabian-Russell Model," International Journal of Hospitality Management, Vol. 8, No. 4, 2009, pp. 494-503.

6
CHAPTER

INVESTIGATING CONSUMER PREFERENCES IN CHOOSING VEGETARIAN RESTAURANTS USING CONJOINT ANALYSIS

Chia-Hsin Cheng[1], Shih-Yen Lin[2*], Chia-Ching Tsai[3]

[1]*Department of Accounting, National Yunlin University of Science and Technology, Yunlin City, Taiwan*

[2]*Department of Hospitality Management, National Chi Nan University, Nanto City, Taiwan*

[3]*Department of Business Administration, National Yunlin University of Science and Technology, Yunlin City, Taiwan*

ABSTRACT

People who eat out in urban are more common than ever before. However, restaurant foods are often high in calories, and lack for enough vegetables. Moreover, they provide too many calories from animal fat and protein resulting in a high risk of cardiovascular and other chronic diseases. Furthermore, a diet high in animal products contributes to global warming. For these reasons, a vegetarian diet can be recommended. The main purposes of this study are the following: (1) Understanding and evaluating the customers' preferences when selecting vegetarian restaurants. (2) In order to decrease the disease risks and the negative impact on the planet caused by a meat diet, researching ways to encourage consumers to choose vegetarian restaurants. The results of conjoint analysis indicated that when people choose vegetarian restaurants, they prefer ones that provide Chinese style cuisines, organic foods, cuisines that can help to lose weight, interiors with simple décor and ones that promote environmental protection.

Keywords: Vegetarian Restaurant, Consumer Preference, Chronic Diseases, Global Warming, Environmental Sustainability, Conjoint Analysis

INTRODUCTION

More and more people work and live in the urban in Taiwan for the convenience who choose eat out every day. According to a survey of Unilever Global Company (2011), the proportion of people that eat out daily in Taiwan has reached 80% of the population; nearly 4.2 times eat in lunch and 3.3 times in dinner away from home per week. However, although eating out is convenient, it is also associated with lower nutritional quality and weight gain.

Increased consumption of away-from-home foods has been associated with the increased intake of calories, saturated fat, added sugars, and sodium, fewer fruits and vegetables, and less fiber, and vitamins (Kant & Graubard, 2004; Schmidt et al., 2005). Furthermore, it is associated with excessive consumption of animal fat and protein, resulting in higher risk of the cardiovascular and chronic diseases. The Department of Health of Taiwan's data show that of the top ten causes of death, seven of them are caused by chronic diseases. Moreover, younger members of society are being diagnosed with these chronic diseases and experts believe it is related to poor diet and nutrition (Lai, 2005). The evidence in Europe show the importance of nutritional factors in preventing disability and death suggesting that 4.4% of the overall burden of disease in the region could be attributed to low vegetarian foods intake (WHO, 2003). Thus, consumption of vegetarian diets such as fruits and vegetables is important for human health because these foods are primary sources of some essential nutrients and contain phytochemicals that may lower the risk of chronic disease (Stahler, 2009). Eating these foods not only can provide rich and high quality vegetable protein and dietary fiber but also can promote intestinal health.

However, in addition to health factors for eating more vegetarian foods, according to Livestock's Long Shadow, the widely-cited 2006 report by the Food and Agriculture Organization of the United Nations (FAO), it estimates that the livestock industry generates 18 percent of annual worldwide GHGs emissions, more than cars and planes GHGs emissions, and these carbon emissions cause serious harm to the climate (Goodland & Anhang, 2009).

Thus, in order to protect the environment, eating more vegetarian diets are actually a responsibility of all people. According the Vegetarian Publication reported that the north city of Belgium, Ghent, has set every Thursday as "Vegetarian Day". This policy began in 2009 and today Ghent is the world's foremost green city. Santiago City of Philippines also passed a similar vegetarian resolution and set every Friday as vegetarian day in the same year (Lai, 2010). From the above, governments are attempting to implement green city plans and to promote vegetarian lifestyles. Eating a more vegetarian diet in addition to enhancing human health is also environmentally friendly.

Conjoint analysis is a very powerful tool for obtaining information about the effect of different product attributes on liking and/or purchase intent of food products (Green & Srinivasan, 1978), and there is little research using conjoint analysis to study consumer behavior on their choices of restaurants. Hence, the main purposes of this study are the following: (1) Understanding and evaluating the customers' preferences when selecting a vegetarian restaurant. (2) In order to decrease the disease risks and the negative impact on the planet caused by a meat diet, researching ways to encourage consumers to choose vegetarian restaurants.

2. LITERATURE REVIEW

2.1 The Importance of Vegetarian Diets

The WHO (2003) has defined health as "not only the absence of a disease or infirmity, but a complete state of physical, mental and social well-being". The recent data indicate that vegetarian diets are important sources of a wide range of vital micronutrients, and there is evidence strongly indicating vegetarian consumption can effectively prevent a number of chronic diseases including cardiovascular disease and some cancers (IARC, 2003; Robertson et al., 2004). Because vegetarian diets contain diverse quantities of antioxidant secondary metabolites that play an important role as health-protective factors, they can neutralize the dangerous effect of free radicals in our cells. Thus, increasing the consumption of vegetarian diets can not only result in the prevention of chronic diseases (Blasa et al., 2010) but also be a very cost-effective method for health promoting (Gundgaard et al., 2003).

On the other hand, according to the Intergovernmental Panel on Climate Change's (IPCC) assessment report "climate change is a result of global warming, and that change is man-made" (Lai, 2010). Goodland and Anhang (2009) indicated the following: "We believe that the supply chain of domesticated animals raised for food have been vastly underestimated as a source of GHGs, the livestock and their byproducts actually account for at least 32,564 million tons of CO_2per year, or 51 percent of annual worldwide GHGs emissions. More than 50% of global greenhouse gases come from livestock, and more than 80% of the Amazon rainforest has been cut down to produce meat, however, the rainforest serves as the lung of the earth and produces oxygen and absorbs carbon dioxide". Moreover, the FAO also pointed out that the phenomenon of over-fishing has led to important fishing grounds in the world being lost. Over-fishing has resulted in the failure of 70% of the most important fish populations (U. N. Report, 2010).

The above-mentioned of U. N. Report claims that a "Vegetarian diet was better for the planet and lesser consumption of animal products is necessary to save the world from the worst impacts of climate change". Therefore, to eat less meat not only can reduce the scale of livestock, but is can also reduce greenhouse gas emissions (Washington, D.C., 2009).

2.2. The Definition and Effects of Product Attributes

Products are items that are valuable for exchange in the process of exchange, and can be exchanged in the market. They can be physical objects or intangible services (Lin, 2002). Each product is made of a combination of many different product attributes, consumers compare competing brands of products and their final choice is based on those product attributes (Cao, 2001). In a word, product attributes are the characteristics of a product. Recently, the population of people eating out is increasingly growing. Consumers are more aware of and care more about nutrition. In the Knutson and Patton (1993) study, they found the mature consumers are increasingly concerned about eating and how it may affect their health and longevity. The WHO (2003) also stated that the impact of vegetarian diets on human health is very helpful. A study of consumers about restaurants' menus found that 21% of respondents thought health and nutrition noted on the menus as important (Glanz et al., 2007). Thus, promoting a healthier diet could be one successful future strategy used by the restaurant industry.

In addition to health reasons, customers typically choose a product or service of a restaurant which consists of multi-attributes (Lewis et al., 1991). Such as food quality, convenience of location; cleanliness; price (Almanza et al., 1994; Knutson & Patton, 1993); hygienic and comfortable environment (Soriano, 2002); restaurant design and layout (Ryu & Jang, 2008); taste and nutritional (Lee & Crange, 2007), moreover, Hu et al. (2010) suggested that green food and green practices of restaurant operations are important factors for consumers when selecting a restaurant. Restaurants that use green products can strengthen their eco-friendly image to attract more customers' attention and enhance customer satisfaction as reported by Manaktola and Jauhari (2007).

Base on above literature mentioned, we conclude that the most important attributes to attract customers include the following: food quality, taste, reasonable price, nutritional aspects, restaurant appearance, less meat can help reduce global warming, clean dining environment, excellent service, well-designed physical facilities, convenient location, and healthy food on the menu. There are many attributes and factors that can influence consumers to choose a restaurant. Restaurant owners can improve their chances of success by being aware of the most important attributes, than planning a healthy menu.

2.3. Conjoint Analysis

The tool chosen for this research is conjoint analysis (CA). In conjoint analysis (CA), the characteristics of a product are described in terms of "attributes". Variations within an attribute are described as "levels". Attribute levels are the values of the independent variables (Reddy et al., 1995). Hence, CA is a multi-variate technique, which determines the relative importance of a product's multi-dimensional attributes and measures the consumers' degree of preferences for each level of each attribute (Green & Srinivasan, 1978). CA allows product attributes to be considered jointly rather than in isolation. Besides, CA can establish the relative importance of different attributes in the provision of a good or a service and also be used to estimate how individuals trade between these attributes (Ryan, 1999). The relative importance of each attribute is calculated as the utility range. CA can better predict the overall consumer preference by aggregating the utility scores of each product attribute. In order to analyze the preference between levels within each attribute, the partial utilities or part-worth have been used, the preferences to determine the value of each attribute (Green & Krieger, 1996). For each level of each attribute, large part-worth values are assigned to the most preferred levels, and small part-worth values are assigned to the least preferred levels. The part worth estimates can serve as a basis for predicting the choice probabilities of various combinations of attribute levels (Myung, 2003).

Usually, a fractional factorial design is used, and uses the orthogonal array in experimental design, which assumes away most interactions among the independent variables considering only main effects (Lewis et al., 1991). The attributes may be related directly to the sensory properties of the product or service provided by the experimenter. For each of the factor combination, each consumer is asked to give his/her liking or purchase intent score (Nñs et al., 2001).

3. RESEARCH METHOD

3.1. Procedure

There are predominantly four stages in the design of a CA study as described by Lehmann et al. (1998), and Van der Pol and Ryan (1996), the design of each of the stages as applicable for our study is discussed below.

3.1.1. Establishing the Attributes to Be Tested

The first stage in the design of a CA study is to establish the attributes to be included in the study. It was recognized that many attributes will have potential influence on an individual's choosing vegetarian restaurants. Based on the above literature mentioned and on interviews with owners of vegetarian restaurants who choose the 8 most important items of vegetarian restaurants preliminary attributes and designed the questionnaires, then, we used the convenience sampling to survey 50 respondents who were eating in the vegetarian restaurants in Taichung from December 3 to December 13, 2013. We request that those respondents choose the 5 most important attributes by ranking those items when searching for a vegetarian restaurant. Order 1 means the most important attribute and order 5 is the least important attribute. Using the data collected from those questionnaires, we used reverse scoring: order 1 is given a score of 5, and order 5 is given a score of 1. Next, we calculated all the scores of attributes, and then summed up the 5 most important attributes which are shown on Table 1. They include the following five attributes: 1) food quality, 2) cuisine style, 3) health concept, 4) restaurant design, and 5) environmental protection. These five attributes were the top five most important ones when selecting vegetarian restaurants.

3.1.2. Assigning Levels to the Attributes

Based on the steps of CA analysis which were proposed by Green and Srinivasan (1978), our study chose two or three levels for each attribute (see Table 2). Levels chosen for each attribute must be defined in a way that is meaningful to respondents in order to ensure the attributes are measured objectively and are able to be controlled (Centre for International Economics, 2001). In this stage, the attribute levels to be used in the experiment were set using both the previous questionnaires information and exploratory interviews carried out with managers who were well acquainted with the area (expert judgment method and group interviews). Then, the five typography design attributes and their corresponding levels were defined, as shown in Table 2.

3.1.3. Establishing Which Scenarios to Present Respondents

There are 48 ($3^1 \times 2^4$) possible profiles (all the possible attribute combinations) to be presented for each use. A very large number of possible vegetarian restaurant attributes would be generated. Presenting many profiles to respondents would result in respondent fatigue and poor results. In order to reduce this complexity and the associated negative impacts, our study used the orthogonal fractional factorial design (Phadke, 1989) reduce the relative important profiles presented to 8 while still allowing for the main effects to be estimated without confounding results. The total number of final stimuli description of the 8 profiles is shown in Table 3.

Table 1. Results of attributes for customers selecting vegetarian restaurants.

Attributes	Total Scores
Food quality	153
Cuisine style	139
Health concept	137
Restaurant design	91
Environmental protection	84
Price	73
Diversity of menu	63
Professional service	17

3.1.4. Establish Preferences

The 8 stimuli are the vegetarian restaurants portfolio to interview respondents, and we used the full profile to present the stimuli data. That means each set of attributes and the level of the portfolio of products with the letter detailed description to design the questionnaire and complete presented to respondents, and investigate the preference of stimuli. Respondents were asked to read the letter description than rate their preference for each scenario from 1 to 8 (1 = most preferred, 8 = least preferred). A carefully constructed introduction preceded the CA section of the interview to allow the respondents to clearly understand the context of the study.

3.1.5. Conjoint Analysis of Pilot Study

In order to confirm the validity of the content of the 8 stimuli of the product portfolio, a pilot study of 50 senior students from the Department of Hospitality Management was undertaken on February 21, 2014 in I-Shou University in Kaohsiung. At this stage, participants were asked whether the items were appropriate for the questionnaires. A total of 50 pilot questionnaires were collected, six questionnaires were incomplete, and thus, the final sample size for the pilot statistical analysis was 44. Pearson's R was .860 and Kendall's tau = .786 in the pilot study which indicates that the overall fit of the model and has good predictability for the product preference of choosing vegetarian restaurants.

Table 2. Attributes and attribute levels.

Typography Design Attributes	Attribute Level in the Research Design
1.Food quality	(1) Organic food
	(2) Colorful food
	(3) Nutritious food
2.Cuisine style	(1) Chinese style
	(2) Italian style
3.Health concept	(1) Lose weight
	(2) Prevent diseases
4.Restaurant design	(1) Simple design
	(2) Fashion design
5.Environmental protection	(1) Care
	(2) Doesn't care

Table 3. Stimuli profile of vegetarian restaurants.

Stimuli Profile	Food Quality	Cuisine Style	Health Concept	Restaurant Design	Environmental Protection
1	Organic	Chinese	Lose weight	Simple	Care
2	Organic	Italian	Prevent diseases	Fashion	Doesn't care
3	Organic	Chinese	Prevent diseases	Fashion	Care
4	Nutritious	Chinese	Lose weight	Fashion	Doesn't care
5	Colorful	Italian	Lose weight	Fashion	Care
6	Colorful	Chinese	Prevent diseases	Simple	Doesn't care
7	Organic	Italian	Lose weight	Simple	Doesn't care
8	Nutritious	Chinese	Prevent diseases	Simple	care

3.2. Survey Instrument

The questionnaire used in this study consisted of three major parts. The first part of the questionnaire comprises 5 items to assess willingness, motivation, consideration, attractiveness and budget when selecting vegetarian restaurants. The second part of the questionnaire is comprised of 8 stimuli of the product portfolio with the letter detailed description of vegetarian restaurants to assess preference ranking: ranking 1 means the first preferred level and ranking 8 means the lowest preferred level. The third part of the questionnaire consists of 7 items to investigate the background information of respondents including sex, age, occupation, education, religious belief, and income.

An on-site survey using self-administered questionnaires was adopted as the method of data collection. It was done at MRT station in Taipei, Taichung and Kaohsiung, from March 10 to April 15, 2014. Our study used random sampling to recruit one of every five customers when they passed by the interviewers. A total of 250 questionnaires were collected; however, 31 questionnaires were incomplete, which resulted in their deletion. The final sample size for statistical analysis was 219. The response rate was 87.6%.

4. RESULTS

4.1. Participants

Table 4 summarizes the demographic profile of the respondents. Respondents consisted of 39.3% male participants and 60.7% female participants. Nearly three quarters (74%) of respondents were between the ages of 20 to 40, having an educational level of Bachelor degree (66.2%). Approximately 32.9% of respondents are engaged in service industry, and 30.5% of respondents had monthly incomes between NT$ 25,001 to NT$ 40,000. The majority of respondents were none religious belief (35.6%). 32.0% of respondents lived in Taipei.

4.2. The Results of Consumer Preference for Selecting Vegetarian Restaurants

Table 5 summarizes the motivation for selecting vegetarian restaurants, the main motivation was "for health" (42.5%), the second motivation was "prefer natural food" (28.8%) and the third motivation was "reducing global warming" (10.0%). Table 6 indicates the first consideration to go to the vegetarian restaurants was "for health" (32.9%) and the second consideration was "food type" (31.5%); as regards the main attractiveness of consuming in a vegetarian restaurant, "for health" (29.2%) again ranked first and "food taste" (25.6%) ranked second. Most of the respondents (42.9%) were willing to pay NT$150 - NT$280 for a vegetarian meal, and 33.3% of respondents were willing to pay NT$281 to NT$400 (See Table 7).

4.3. Results of Conjoint Analysis

Conjoint analysis was used to analyze the preference of product attributes for consumers who choose vegetarian restaurants including food quality, cuisine style, health concept, restaurant design and environmental protection. Respondents were asked to rate their preference for each scenario from 1 to 8 (1 = most preferred, 8 = least preferred).

Table 4. Demographics of respondents.

Characteristics	N	% (219)	Characteristics	N	% (219)
Willing			Monthly Income (NT$)		
Yes	205	93.6	Under 10,000	50	22.8
No	14	0.6	10,000 - 25,000	34	16.0
Gender			25,001 - 40,000	67	30.5
Male	86	39.3	40,001 - 55,000	34	15.6
Female	133	60.7	55,000 - 70,000	16	7.3
Age			More than 70,001	17	7.8
20 or younger	9	4.1	Religious Belief		
21 - 30	105	47.9	Buddhism	61	27.9
31 - 40	57	26.0	Taoism	62	28.3
41 - 50	39	17.8	Christianity	6	2.7
51 or older	9	4.1	Catholicism	2	0.9
Education			None	78	35.6
Junior	3	1.4	Other	10	4.6
Senior	13	5.9	Location		
Bachelor	145	66.2	Taipei	70	32.0
Master or above	58	26.5	North of Miaoli	20	9.1
Occupation			Taichung	39	17.8
Students	58	26.5	Yulin &Jiayi	21	9.6
Service	72	32.9	Kaohsiung	32	14.6
Manufacturing	16	7.3	Pingtung	32	14.6
Science	5	2.3	East of Taiwan	1	0.5
Finance	13	5.9	Other	4	1.8
Teaching	28	12.8			
Free	7	3.2			
Unemployed	6	2.7			
Other	14	6.4			

Table 8 shows that the relative importance values and the favorite attribute levels of the consumers selecting vegetarian restaurants. We found that Pearson's R was .820 and Kendall's tau was .771 which indicated that the overall fit of the model was good. The results confirm that conjoint analysis has good predictability for consumers choosing vegetarian restaurants. "Food quality" is the most preferred factor with a relative importance value of 34.62% in selecting vegetarian restaurants, followed by "environmental protection "(17.27%). As the food quality aspect, consumers prefer the

Table 5. Motivation of participants of choosing vegetarian restaurants.

Characteristics	N	% (219)
Motivation		
For health	93	42.5
Prefer natural food	63	28.8
Reducing global warming	22	10.0
Other	20	9.1
Religious belief	11	5.0
Vogue	10	4.6

Table 6. Consideration and Attractiveness for choosing vegetarian restaurants.

Consideration	N	% (219)
For health	72	32.9
Food type	69	31.5
Food taste	42	19.2
Décor & atmosphere	13	5.9
Reducing global warming	12	5.5
Green restaurant	9	4.1
Other	2	0.9
Attractiveness		
For health	64	29.2
Food taste	56	25.6
Food type	52	23.7
Décor & atmosphere	22	10.0
Green restaurant	11	5.0
Reducing global warming	8	3.7
Other	6	2.7

vegetarian restaurants which provide "organic food" (.344); as the cuisine style aspect, consumers prefer "Chinese style"(.064); about the health concept aspect, consumers prefer "lose weight" (.048); related to restaurant design, respondents prefer a "simple design"(.023), with regard to the environmental protection aspect, respondents prefer vegetarian restaurant owners whose mission of operation is to "care" about environmental protection and sustainability (.233).

Table 7. Budget of participants for choosing vegetarian restaurants.

Characteristics	N	%(219)
Willing to Pay		
Under NT $150	28	12.8
NT $151 ~ NT $280	94	42.9
NT $281 ~ NT $400	73	33.3
NT $401 ~ NT $500	20	9.1
More than NT $551	4	1.8

5. DISCUSSION

This study investigated the preferences of customers selecting vegetarian restaurants. The results indicate that 93.6% of consumers are willing to dine at vegetarian restaurants, and most consumers who choose a vegetarian restaurant are females. The main ages for those choosing vegetarian restaurants are adults between 21 to 40 (See Table 4). This result is similar to Perry's (2001) study which found that younger females (under 35) that chose vegetarian restaurant and their primary reason were concern about weight. They believed they could lose or control weight by eating a vegetarian diet as compared to a non-vegetarian one.

From the results of conjoint analysis indicated "food quality" can be the contributing factor when consumers choose vegetarian restaurants. These results are consistent with the findings of previous studies (Auty, 1992; Knutson & Patton, 1993; Almanza et al., 1994; Sun, 1995; Soriano, 2002; Lo & Lam, 2004; Sulek & Hensley, 2004), which disclosed that food and beverage quality is the most important factor during the restaurants selection process, because people believe that dining pleasure and satisfaction comes from high quality food and beverages (Turner & Collison, 1988). The investigation shows that "organic food " is the most important factor for customers when assessing food quality, because most consumers not only thought that organic food was considered healthier than conventional food (Grankvist & Biel, 2001; Makatouni, 2002; Zanoli & Naspetti, 2002;

Table 8. Aggregate-level results of conjoint analysis.

Typography Design Parameters	Attribute Utility	Relative Importance	Preference
Food Quality		34.62%	
(1) Organic food	.344		V
(2) Colorful food	−.215		
(3) Nutritious food	−.129		
Cuisine Style		16.85%	
(1) Chinese style	.064		V
(2) Italian style	−.064		
Health Concept		15.30%	
(1) Lose weight	.048		V
(2) Prevent disease	−.048		
Restaurant Design		15.95%	
(1) Simple design	.023		V
(2) Fashion design	−.023		
Environmental Protection		17.27%	
(1) Care	.233		V
(2) Doesn't Care	−.233		

Pearson's R = .820 Sig. = .0063. Kendall's tau = .771Sig. = .0239.

Radman, 2005), but felt that was much more beneficial to eat food grown without chemical additives, and the taste of organic food was considered by some to be superior to that of conventional food (Lockie et al., 2002; McEachern & McClean, 2002; Radman, 2005). On the other hand, they think organic food is associated with health and social concerns as well as economic and environmental sustainability (Poulstona & Yiua, 2011).

This study investigated the second most relatively important factor for selecting vegetarian restaurants which was "environmental protection". The results were compatible with previous studies (Manaktola & Jauhari, 2007; Hu et al., 2010) which indicated that the green restaurant are an important factor for consumers when selecting a restaurant. Because consumers believe that green products/practices are beneficial to the environment (Kalof et al., 1999) including eating less meat can get result in energy savings and carbon reduction, therefore, when firms use green products/practices, these products/practices not only enhance the product or company's eco-friendly image, but also enhance customer satisfaction. Such customers are willing to pay more for environmentally friendly products/services (Manaktola & Jauhari, 2007; Vieregge et al., 2007).

"Cuisine style" was the third ranked relatively important factor of the conjoint analysis when consumers were selecting vegetarian restaurants. The finding is consistent with Koo et al. (1999), Liu and Jang (2009) who reported the style of cuisine is the important attribute when selecting a restaurant. Most of these consumers preferred "Chinese style" cuisine. In general, people prefer their own native flavors, owing to "Chinese food and beverage culture" it has been developed for more than 5000 years. And Chinese cuisine is prepared over vast areas with great regional diversity. Therefore, "Chinese style" cuisine is the preferred taste for consumers in our study. The study shows that restaurant design is the fourth relative factor when selecting vegetarian restaurants. More and more consumers believe that when restaurants make efforts on designs, layouts, exterior and interior decorations, these efforts will enhance their pleasure and satisfaction of the dining experience (Ryu & Jang, 2008). Most consumers in our sample preferred a "simple design". That finding appears to be consistent with previous studies (Brady & Cronin, 2001; Raajpoot, 2002). This finding shows that a simple but comfortable environment atmosphere will not only enhance the pleasure and satisfaction of the dining experience but also will play a specific role in customer decision making.

The "health concept" ranked as the fifth most important factor when selecting vegetarian restaurants. Consumers preferred the "lose weight" concept and the findings are consistent with the study of Knutson and Patton (1993) that found that many mature consumers are increasingly concerned about health diet can help to lose weight. Most people are aware that they need to increase their fiber intake, fruits and more vegetarian foods in their diet to help them decrease chronic diseases (Wasserman & Stahler, 1998) and promote their health.

6. CONCLUSIONS

The findings of this study indicate that Taiwanese who choose vegetarian restaurants prefer to lose weight, high food quality and food that is organic. They also prefer Chinese style cuisine especially cuisine method can help to lose weight, and restaurants that offer a simple design atmosphere, as well as these restaurants care about environmental protection. The most important motivator and main consideration as well as attractiveness were "for health", and thought that the "food quality" is the relatively important factor of other vegetarian restaurants attributes. The U.S. National Cancer Institute has initiated the " 5 A Day for Better Health" plan to encourage adults to consume at least 5 servings of F & Vs daily (CDC, 2007). In March 2007, the " 5 A Day" goal was replaced with 7 - 13 (3 1/2 - 6 1/2 cups) servings F&Vs daily (USDHHS, 2010). Owing to lifestyle changes, more and more people are concerned with health and longevity. Hence, they recognize that good food and nutrition can improve health and disease prevention, and in general, vegetarian foods have higher amounts of fiber, vitamins E, A, and C, thiamin, riboflavin, calcium, magnesium, iron, and potassium, and those ingredients not only prevent chronic diseases, such as obesity, hypertension, coronary disease, breast and colon cancer, but also are good for losing weight. In Wasserman and Stahler' findings (1998), they reported that 82% of a sample of adults in the United States chose to eat a vegetarian diet for health reasons.

Moreover, for ethical and environmental reasons, eating a vegetarian diet has been shown to contribute to "ecosystem health" by reducing the impact on the environment and economies of pollution, and land degradation by grazing (Fox, 1999). A vegetarian diet saves water, has a smaller impact on global warming, decreases deforestation, is more humane with regards to animals, and protects the soil. Global warming is now recognized as a serious problem. Research has shown that a vegetarian diet is environmentally friendly and one way global citizens can actively be responsible for improving the situation.

Conjoint analysis is frequently used by marketers to understand consumer attitudes about a product with new attributes or features. A better understanding of people's reasons for choosing vegetarian restaurants would help vegetarian restaurant owners to design the most preferred product by customers and for planning products as well as make better decisions including providing nutritious menus, designing marketing strategies to promote their menus. These steps might encourage more people to dine frequently at vegetarian restaurants. Simultaneously, we hope the results of this study were expected to playing an important role in solving some of the world's most challenging environmental problems.

REFERENCES

1. Almanza, B. A., Jaffe, W., & Lin, L. (1994). Use of the Service Attribute Matrix to Measure Consumer Satisfaction. Journal of Hospitality & Tourism Research, 17, 63-75.

2. Auty, S. (1992). Consumer Choice and Segmentation in the Restaurant Industry. The Service Industries Journal, 12, 324-339.

3. Blasa, M., Gennari, L., Angelino, D., & Ninfali, P. (2010). Fruit and Vegetable Antioxidants in Health. In Bioactive Foods in Promoting Health: Fruits and Vegetables (pp. 37-58). San Diego, CA: Academic Press.

4. Brady, M. K., & Cronin, J. J. (2001). Some New Thoughts on Conceptualizing Perceived Service Quality: A Hierarchical Approach. Journal of Marketing, 65, 34-49.

5. Cao, S. X. (2001). Tourism Marketing. Taipei: Yang-Chih Book Co., Ltd.

6. Centers for Disease Control and Prevention (CDC) (2007). Fruit and Vegetable Consumption among Adults, United States, 2005 (pp. 213-217). Morbidity and Mortality Weekly Report.

7. Centre for International Economics. (2001). Review of willingness-to-pay methodologies. Prepared for the Independent Pricing and Regulatory Tribunal of NSW, Canberra.

8. Fox, A. M. (1999). The Contribution of Vegetarianism to Ecosystem Health. Ecosystem Health, 5, 70-74.

9. Glanz, K., Resnicow, K., Seymour, J., Hoy, K., Stewart, H., L. M., & Goldberg, J. (2007). How Major Restaurant Chains Plan Their Menus: The Role of Profit, Demand, and Health. American Journal of Preventive Medicine, 32, 383-388.

10. Goodland, R., & Anhang, J. (2009). Livestock and Climate Change.

11. Grankvist, G., & Biel, A. (2001). The Importance of Beliefs and Purchase Criteria in the Choice of Eco-Labeled Food Products. Journal of Environmental Psychology, 21, 405-410.

12. Green, P. E., & Krieger, A.M. (1996). Individualized Hybrid Models for Conjoint Analysis. Management Science, 42, 850- 867.

13. Green, P. E., & Srinivasan, V. (1978). Conjoint Analysis in Consumer Research: Issues and Outlook. Journal of Consumer Research, 5, 103-123.

14. Gundgaard, J., Nielsen, J. N., Olsen, J., & Sørensen, J. (2003). Increased Intake of Fruit and Vegetables: Estimation of Impact in Terms of Life Expectancy and Healthcare Costs. Public Health Nutrition, 6, 25-30.

15. Hu, H. H., Parsa, H. G., & Self, J. (2010). The Dynamics of Green Restaurant Patronage. Cornell Hospitality Quarterly, 51, 344-362.

16. IARC (2003). Fruit and Vegetables Protective Overall against Cancer. International Agency for Research on Cancer, Lyon.

17. Kalof, L., Dietz, T., Stern, P. C., & Guagnano, G. A. (1999). Social Psychological and Structural Influences on Vegetarian Beliefs. Rural Sociology, 64, 500-511.

18. Kant, A. K., & Graubard, B. I. (2004). Eating out in America, 1987-2000: Trends and Nutritional Correlates. Preventive Me- dicine, 38, 243-249.

19. Knutson, B. J., & Patton, M. E. (1993). Restaurants Can Find Gold among Silver Hair. Journal of Hospitality and Leisure Marketing, 1, 79-90.

20. Koo, L. C., Tao, K. C., & Yeung, J. H. C. (1999). Preferential Segmentation of Restaurant Attributes through Conjoint Ana- lysis. International Journal of Contemporary Hospitality Management, 11, 242-253.

21. Lai, C. Y. (2005). The Effects of College Students' Health Diets Behaviors. Master's Thesis, Taipei: National Taiwan Normal University.

22. Lai, J. Y. (2010). Reduce Carbon and Eat Less Meat to Save the Earth. Vegetarian Publication.

23. Lee, S. J., & Crange, D. A. (2007). The Relative Importance of Menu Attributes at Point of Menu Selection through Conjoint Analysis: Focused on Adolescents. Journal of Foodservice Business Research, 10, 3-18.

24. Lehmann, D. R., Gupta, S., & Steckel, J. H. (1998). Marketing Research. Boston, MA: Addison Wesley.

25. Lewis, R. C., Ding, S., & Geschke, U. (1991). Using Trade-Off Analysis to Measure Consumer Choices: The Full Profile Method. Journal of Hospitality &Tourism Research Journal, 15, 75-92.

26. Lin , J. H. (2002). Marketing Management. Taipei: Best Wise Co., Ltd.

27. Liu, Y., & Jang, S. (2009). Perceptions of Chinese Restaurants in the U.S.: What Affects Customer Satisfaction and Behavioral Intentions? International Journal of Hospitality Management, 28, 338-348.

28. Lo, A., & Lam, T. (2004). Long-Haul and Short-Haul Outbound All-Inclusive Package Tours. Asia Pacific Journal of Tourism Research, 9, 161-176.

29. Lockie, S., Lyons, K., Lawrence, G., & Mummery, K. (2002). Eating "Green": Motivations behind Organic Food Consumption in Australia. Sociologia Ruralis, 42, 23-40.

30. Makatouni, A. (2002). What Motivates Consumers to Buy Organic Food in the UK? Results from a Qualitative Study. British Food Journal, 104, 345-352.

31. Manaktola, K., & Jauhari, V. (2007). Exploring Consumer Attitude and Behavior towards Green Practices in the Lodging Industry in

India. International Journal of Contemporary Hospitality Management, 19, 364-377.

32. McEachern, M.G., & McClean, P. (2002). Organic Purchasing Motivations and Attitudes: Are They Ethical? International Journal of Consumer Studies, 26, 85-92.

33. Myung, R. (2003). Conjoint Analysis as a New Methodology for Korean Typography Guideline in Web Environment. Inter- national Journal of Industrial Ergonomics, 32, 341-348.

34. Nñs, T., Kubberød, E., & Sivertsen, H. (2001). Identifying and Interpreting Market Segments Using Conjoint Analysis. Food Quality and Preference, 12, 133-143.

35. Perry, C. L., Mcguire, M. T., Neumark-Sztainer, D., & Story, M. (2001). Characteristics of Vegetarian Adolescents in a Mul- tiethnic Urban Population. Journal of Adolescent Health, 29, 406-416.

36. Phadke, M. S. (1989). Quality Engineering Using Robust Design. Upper Saddle River, NJ: PTR Prentice-Hall.

37. Poulston, J., & Yiu, A. Y. K. (2011). Profit or Principles: Why Do Restaurants Serve Organic Food? International Journal of Hospitality Management, 30, 184-191.

38. Raajpoot, N. A. (2002). TANGSERV: A Multiple Item Scale for Measuring Tangible Quality in Food Service Industry. Journal of Foodservice Business Research, 5, 109-127.

39. Radman, M. (2005). Consumer Consumption and Perception of Organic Products in Croatia. British Food Journal, 107, 263- 273.

40. Reddy, V. S., Bush, R. J., & Roudik, R. (1995). A Market-Oriented Approach to Maximizing Product Benefits: Cases in U.S. Forest Products Industries. US Forest Service, 19-38.

41. Robertson, A., Tirado, C., Lobstein, T., Jermini, M., Knai, C., Jensen, J. et al. (2004). Food and Health in Europe: A New Basis for Action. European Series, No. 96. Copenhagen: WHO Regional Publications.

42. Ryan, M. (1999). Using Conjoint Analysis to Take Account of Patient Preferences and Go Beyond Health Outcomes: An Application to in Vitro Fertilization. Social Science and Medicine, 48, 535-546.

43. Ryu, K., & Jang, S. (2008). DINESCAPE: A Scale for Customers' Perception of Dining Environments. Journal of Foodservice Business Research, 11, 2-22.

44. Schmidt, M., Affenito, S. G., Striegel-Moore, R., Khoury, P. R., Barton, B., Crawford, P. et al. (2005). Fast-Food Intake and Diet Quality in Black and White Girls: the National Heart, Lung, and Blood Institute Growth and Health Study. Archives of Pediatrics & Adolescent Medicine, 159, 626-631.

45. Soriano, D. R. (2002). Customers' Expectations Factors in Restaurants: The Situation in Spain. International Journal of Quality and Reliability Management, 19, 1055-1068.

46. Stahler, C. (2009). How Many Vegetarians Are There? Vegetarian Journal, 4, 6-9.

47. Sulek, J. M., & Hensley, R. L. (2004). The Relative Importance of Food, Atmosphere, and Fairness of Wait: The Case of a Full-Service Restaurant. Cornell Hotel and Restaurant Administration Quarterly, 45, 235-248.

48. Sun, L. H. (1995). Consumer Involvement in Restaurant Selection: A Measure of Satisfaction/Dissatisfaction (I). Journal of Nutrition in Recipe and Menu Development, 1, 45-58.

49. Turner, M., & Collison, R. (1988). Consumer Acceptance of Meals and Meal Components. Food Quality and Preference, 1, 21-24.

50. U. N. Report (2010). UN Urges Global Move to Meat and Dairy-Free Diet. Lesser Consumption of Animal Products Is Necessary to Save the World from the Worst Impacts of Climate Change.

51. U. S. Department of Health and Human Services (USDHHS) (2010). Dietary Guidelines for Americans.

52. Unilever Global Company (2011). The Population for Eating out in Taiwan.

53. Van der Pol, M., & Ryan, M. (1996). Using Conjoint Analysis to Establish Consumer Preferences for Fruit and Vegetables. British Food Journal, 98, 5-12.

54. Vieregge, M., Scanlon, N., & Huss, J. (2007). Marketing Locally Grown Food Products in Globally Branded Restaurants: Do Customers Care? Journal of Foodservice Business Research, 10, 67-82.

55. Washington DC, Word Watch (2009). Livestock Emissions: Still Grossly Underestimated. November/December.

56. Wasserman, D., & Stahler, C. (1998). Reader Survey Results. Vegetarian Journal, 17, 4.

57. WHO (2003). Definition of Health.

58. Zanoli, R., & Naspetti, S. (2002). Consumer Motivations in the Purchase of Organic Food: A Means-End Approach. British Food Journal, 104, 643-653.

7 CHAPTER

INNOVATIVE FOOD SAFETY STRATEGIES IN A PIONEERING HOTEL

Su-Ling Wu

Department of Leisure Management, Yu Da University, Miaoli County, Chinese Taipei.

ABSTRACT

A case study approach was adopted to identify the innovative food safety strategies in place at one pioneering hotel that had voluntarily implemented food safety control. An investigation of food safety strategies and the reasons for their implementation in the hotel foodservice were carried out using multiple sources of data, including interviews with key decision makers in the hotel, observations of the business environment, and a review of documentation. The findings suggest that not only food control strategies but also marketing and corporate strategies are important when addressing the problems of food safety. The findings of this study also demonstrate the relationship between motivating factors and food safety strategies. Analysis of the interviews indicates that a hotel's food safety strategies depend significantly on the attitude of senior management, the firm's capability, and corporate image.

Keywords: Food Safety Strategies; Food Safety Control System (FSCS); Hotel

1. INTRODUCTION

Despite the advances in technology and hygiene standards in the past three decades, the outbreaks of fatal food born illness continued to increase globally, such as outbreaks of Salmonella serotype Enteritidis (SE), Listeria monocytogenes (Lm) and Bovine Spongiform Encephalopathy (BSE) in Europe and Northern America; Escherichia coli O157:H7 in Japan; cholera in Asia, Africa, and Latin America [1]. Life style changes and rapid global economic growth contributed to this trend [1]. These two factors lead to great numbers of people eating processed or imported food, and consuming food in foodservice establishments; and the increase of global movement in food supply and people also increase the exposure rate of contacting food born illness due to the unawareness of food born hazard on imported food or on food consumption while traveling aboard [1]. To help curb the rate of food born illness generated by the food industry, government agencies in Taiwan and other countries continue to pass new regulations, such as mandatory or voluntary enforcements of Good Manufacturing Practices (GMP), Good Hygienic Practices (GHP), Hazard Analysis Critical Control Points (HACCP) and ISO 9000 and 22000 series. Moreover, other stakeholders, such as food safety advocacy groups, customers, the press and media, and others, are pushing for firms to take food safety strategies from farm-to-fork, such as risk assessments throughout the external supply chains and the internal production process proactively into their managerial decision making process to minimize contamination which lead to loss of reputation, revenue, and even disruption of business. The incidence of food borne illness may damage the reputation of a country as a tourist destination [2]. The dining experiences of tourists were among the most important predictors of overall travel satisfaction [3,4]. Other researches indicated that the perception of food risks at their travel destinations also influenced their purchasing decisions for other products, especially when they were traveling to less developed countries [5,6]. Therefore, implementation of a certified food safety standard such as the GHP and HACCP standards would benefit especially less developed countries or regions, like Taiwan, by offering food quality assurance, strengthening the foodservices' competitiveness and ultimately enhance the country or region's reputation as tourist destination. In Taiwan approximately 40% of food borne illness outbreaks was linked to commercial food establishments (http://www. fda.gov.tw). Due to these repeated outbreaks of food borne illness, the Taiwanese Health Department in 2000 published a Food Safety Control System (FSCS), which includes principles from both HACCP and Good Hygienic Practice (GHP), in order to bring Taiwan in line with international standards. Beginning in September 2008, the Taiwanese Heath Department has been promoting the FSCS standards with stringent guidelines for the hotel industry. These standards have been gradually taking effect in outlets with central kitchens or kitchens in international tourist hotels. In order to promote the Taiwanese tourism industry, since 2009 the Taiwan Tourism Bureau has provided financial incentives to encourage the hotel industry to improve their facilities and service quality to meet international standards by obtaining local or internationally recognized certificates such as FSCS (http://admin.taiwan.net.tw).

Consumers react strongly to negative reports of food safety and their purchasing behaviors are negatively influenced [7]. Consumer food safety concerns have increased rapidly in recent years, as several high profile food incidents grabbed the head lines. Nevertheless serious incidents continue to occur. Food safety concerns have become an important factor for consumers in determining food selection and consumption. Chemical food additives, meats from diseased pigs and pesticide residuals are considered to be the most severe food safety problems by consumer in Taiwan, especially while they dine in commercial food outlets and consume processed food or meat products [8]. Similar results were also found in two studies which indicated that food safety issues influenced consumers' dining-out decisions [9,10]. Fears of being featured on the media and press were one of reasons for foodservice initiating food safety control program [11]. Therefore, the foodservice industry should try to build their reputation by offering less pre-packaged, processed food and more on healthy, nutritious fresh certified produces. Food Safety Control System (FSCS) specifies the requirements needed for food safety management to ensure the final food is safe for consumption. Therefore, the innovative food safety strategies currently adopted and the strategies being carried out in a pioneering hotel are the focus of this study.

Previous hospitality food safety research has focused only on the food safety control strategy. However, that the levels and types of food safety control strategy include not only a functional strategy (control and marketing strategy), but also a business (competitive advantage) and a corporate strategy (a link with quality and other corporate objectives) [12,13]. The main aim of this study was to understand the food safety strategies (activities) in a hotel food service, while government bureaus have been gradually putting requirements and standards in place, and encouraging tourist hotels to gain FSCS certification. This study explores experiences of the implementation and operation of FSCS in a pioneering international tourist hotel in Taiwan, and thus identifies the key motivating factors and the challenges to implementing food safety strategies.

2. METHOD

An investigation of food safety strategies and the reasons food safety strategies are or are not implemented in hotel food services was carried out using multiple sources of data, including interviews, observation of environmental conditions, and the review of documentation to increase the construct validity and reliability of this study [14]. The method was designed to yield results that would assist the prediction of future drivers of food safety strategies and their uptake in the hotel food service industry. A franchised international tourist hotel that had voluntarily implemented the Food Safety Control System (FSCS) was identified as a suitable participant for the purposes of the study.

Semi-structured interviews were conducted with senior management in the hotel, such as managers, chefs, and food safety supervisors, who were in management teams that had received special training in food safety control system principles [13]. The interviewer (the researcher) specifically directed open-ended questions to the interviewee in relation to food safety strategies. A set of questions was as follows:

1) What are the food safety practices in your company? How are they carried out in your firm? Why?

2) Is it convenient for you to practice food safety strategies? If not, how could it be made convenient for you?

3) What is your biggest food safety management difficulty?

4) Is there anything that would assist you in your food safety management efforts?

With the consent of the participants, interviews were taped and later transcribed. A pattern matching analytical procedure was used to increase the internal validity of a study [14]. It was anticipated that this analysis would lead to a set of food safety strategies being identified and their motivators.

The case study was conducted in a hotel that was a locally owned franchise of an international hotel chain. The chain's first hotel opened in the early 1990s in Taiwan and there are now 9 hotels in the chain. The company has been striving to reinforce food safety strategies voluntary since early 1999 in all its hotels. It is company policy that all hotels in the chain adopt a food safety control system (FSCS). The hotel that participated in this study obtained FSCS certification in 2011.

3. RESULTS AND DISCUSSION

Seven interviews with staff from the chain hotel were conducted for the case study. The interviewees included managers in the food and beverage departments, executive chefs, food safety supervisors, personnel managers and general managers. Most had been in the hotel industry for more than 10 years. All of them had received training in food safety and received FSCS certification. The average time per interview was 35 minutes. Observations of environmental conditions and examination of documentation such as annual reports, policy statements, employee manuals, product advertisements, and corporate advertisements were also collected in order to obtain a greater understanding of operational systems. Many interesting issues were found during analysis of the interviews. The interview data revealed that the study hotel engaged in many food safety strategies, such as safety control, marketing, and corporate organization.

3.1. Safety Control Strategies

The study hotel's capacity to provide sufficient personnel, a training program, time, and resources was found to be the main barrier to the consistent use of a food safety control system [15, 16]. As far as possible, however, the principles of the FSCS were adhered to by the various departments of the hotel, as well as the hotel's suppliers. Production equipment; cleaning and sanitation; personal hygiene; training; chemical control; receiving, storage, and shipping; traceability and recall; and pest control (http://food.fda.gov.tw) were all subject to provisions outlined in the FSCS. The interviewees indicated that the hotel's Food Safety Control System (FSCS) could be successfully carried out with proper layout and equipment and an effective training and assessment program.

3.1.1. Layout and Equipment

Analysis of the data from the interviews revealed that proper layout and equipment in the kitchen promoted food safety and supported food safety systems. With a more centralized kitchen, the interviewees indicated that less cost, better food quality, and less food borne risk could be achieved [13].

"Our hotel has a centralized kitchen in the basement of the building where food ingredients are pre-processed and supplied to all the kitchens in the hotel. Cost can be reduced through easier control in inventory, better deal in purchasing with large quantities, and less labor and equipment needed in each kitchen. Food quality is easier to control, not only in food freshness, texture, and portion size, but also in food safety. Different areas for different kinds of food in process and storage are designed to eliminate cross contamination. Examples of physical separation include the space of food preparation and refrigeration separated according to the food type such as seafood, meat, cold process food, hot process food, and bakery goods." (Food and Beverage Manager, 20/07/2010, interview at the hotel office).

3.1.2. In-House Training and Assessment

Effective food safety training should be designed to closely match employees' expectations and the organization's goals and be accompanied by inducements that motivate staff in order to optimize training performance [13,17,18].

"In Taiwan, kitchen staffs need to have eight hours food safety seminar attendance each year or 32 hours in four years to keep chef certification valid. FSCS certified kitchen needs to have 80% of kitchen staffs with chef certifications. In addition, at least three staffs need to have additional 32 hours FSCS training course and pass qualify exam at the end of the course in order to form a FSCS team." (Executive chef, 20/07/2010, interview at the hotel office).

For the food safety seminar, overall, the interviewees agreed that effectiveness of training is an important factor in gaining the employees' commitment to the company's goals in terms of the FSCS. However, all the interviewees agreed that the current training courses and certification awarded by the HACCP associations should offer a balance of practical skills development and knowledge requirements [13]. Proper food safety handling skill should also be included in training courses and assessed by both theory and a practical exam. Overall, the interviewees indicated that hotel staff who attended the HACCP courses all felt that the information sessions were overlong long and that the content was boring.

"The idea that the HACCP course focuses on both GHP and HACCP plan development seems ok theoretically. However, it is hard for our staffs to sit and listen eight hours a day. An interactive short course should be developed by incorporating more case discussion after showing real examples of HACCP operation in pictures or video rather than too much written words theory. In addition, practical skill development on food preparation should also be included." (Personnel Manager, 20/07/2010, interview at the hotel office).

An effective training system should not be without effective food safety management and compliance assessment. The interviewees indicated that hotel food safety was supervised and audited by an assistant manager in the Stewarding department who worked closely with the executive chef, all restaurants and kitchens, and was fully empowered by the General Manager to communicate across departments on the issue of food safety. The General Manager of the hotel was very conscious of the food safety issue, and had considerable experience in the operation of hotel kitchens and FSCS implementation. He always checked and made corrections on the food safety operation while walking around the hotel, and often walked with the food safety supervisor. A video recording system was also used to assist the assessment of activities in the hotel.

"Our hotel has video camera around the food process area, and the food safety auditor will review the video to see any defaults during operation. The auditor will also walk around the kitchen floor, take pictures, and make correction. The audit report and video will be discussed with the kitchen staff during weekly meeting in which the staff will be able to identify the problem from practical examples. It is also hope to increase the staff involvement through the weekly discussion. The practical examples can also be a good material for food safety training." (Food safety auditor, 20/07/2010, interview at the hotel office).

The Area Manager from the franchisor's head office also conducted spot-checks of food safety standards implementation. In addition, every restaurant in the hotel formed an audit team, which then assessed the other restaurants on a monthly rotation, and reported to the executive chef. The Executive Chef in an interview indicated that temperature measuring and recording were especially important for FSCS implementation, but that not all the chefs were able to do the measuring and recording work. However, the senior management ensured that they received assistance.

"Many experienced chefs, especially in Chinese cuisine, usually have lower levels of education. Some of them have difficulty to read and write and do not have hospitality relevant degrees. Therefore, now we have kitchen stewards to assist the chef on measuring and record keeping to overcome this problem." (Executive chef, 20/07/2010, interview at the hotel office).

As for environmental issues, many interviewees indicated that wearing disposable gloves was not necessary because, unlike the circumstances in a factory, most of their products were ready to eat after a short processing time. They noted that the bacteria on hands after proper washing was not enough to cause sickness, while the disposable gloves not only caused environmental problems, but also allowed employees to forget to actually wash their hands.

3.2. Marketing Strategies

Many studies [9,10,19,20] have shown that awareness of food safety or having the experience of food poisoning did influence consumers' dine out decisions.

3.2.1. Transparent/Open Kitchen

The interviewees indicated that the customers' trust concerning food safety and quality maintained or improved the company's brand image [13]. The presentation of transparent and open kitchens staffed with employees well trained in food safety was a sign to customers that food was being prepared in a safe way and that there was nothing to hide.

3.2.2. Advertising

The hotel advertised their food safety efforts by posting where their food came from and when they received FSCS certification on both their hotel's webpage and restaurant entrances.

3.2.3. In-House Training for the Public

In one interview, a General Manager indicated that the hotel held in-house FSCS training courses sponsored by one of the HACCP associations, and the training was not limited to their hotel staff, but that others could participate.

"At the end of the course, the trainee will take turn to observe and assess our Hotel FSCS system, then form a discussion after that. It is another way to advertising our hotel by having this activity." (General Manager, 20/07/ 2010, interview at the hotel office).

The interviewee indicated that they hoped the customers and the trainee could judge the positive nature of the food safety effort from the dining environment and staff presentation.

3.3. Corporate Strategies

3.3.1. Corporate Social Responsibility

In this study addressing food safety issues was found to be part of the company's policy and was documented on the mission and vision statements of the company. All the chain hotels were required to obtain food safety control system (FSCS) certification. The interviewees agreed that protecting the health and safety of the customers and employees by providing a healthy and safe environment was part of the hotel's corporate social responsibility (CSR). Many food safety efforts made by the company are outlined in annual reports and highlighted on the hotel webpage.

3.3.2. Management Team

A FSCS management team was led by the hotel General Manager and consisted of the executive committees and other heads of department, including the Food and Beverage Manager, Chief Engineer Controller, Executive Chef, Sales and Marketing Manager, Personnel Manager, Restaurant manager, Banquet Manager, Stewards, Bar Manager. The FSCS team met monthly to discuss data gathered during assessments of the hotel, and areas needing further improvement. The General Manager indicated that FSCS was a task that needed all departments to work together in order to effectively achieve required standards of food safety.

3.4. Food Safety Strategies and Their Motivators

Analysis of the interviews indicates that the commitment of senior management to food safety, the firm's capacity to act, and ideas of competitive advantage have the most positive effect on overall food safety strategies. Table 1 shows extracts of significant food safety strategies and primary motivators from the interview data.

Table 1. Thematic conceptual matrix-food safety strategies and motivator.

	Food safety Strategies	Motivators
Safety Control Strategies	Layout & equipment In-house training/assessment	Senior management, firm's capability, Senior management, firm's capability
Marketing Strategies	Transparent/open kitchen Advertisement In-house training for public	Corp image, product differentiation, Corp image, Corp image
Corporate Strategies	CSR management team	Corp image, Senior management

The commitment of the management of a hotel to standards of food safety is the key driving factor for making the resources available for the implementation of a food safety control system [13]. In Asia, most corporations still operate on a traditional hierarchical basis in which decisions are made from the top [21]; therefore, commitment on the part of senior management is especially important. The implementation of a food safety control system in the hotel was mainly influenced by the senior management, who realized that FSCS needed the cooperation of all departments. The interviewees also indicated that a firm's willingness to support FSCS relied on the proper provision of facilities and an in-house food safety training program with incentives to retain and motivate employees were significant motivators in promoting food safety program.

The objectives of competitive advantage are cost reduction and differentiation [22], such as developing sound safe food products to improve reputation and brand loyalty. As mentioned earlier, cost saving can be easily achieved, as well as food quality and safety, with a centralized kitchen. However, interviewees indicated that the overall cost saving was minimal, compared to the investment cost. The interviewees stated that the senior management all realized the international trend toward food safety management and were willing to invest.

The franchising company's policy was to have all the hotels in the chain obtain food safety control system (FSCS) certification in order to create a brand image which customers could trust in terms of food safety and quality. Although the system was mostly carried out behind the centralized kitchen, transparent and open kitchen facilities in the dining areas were built for customers to see the hotel's product differentiation and efforts to achieve food safety and quality. Advertisement on the hotel webpage and in-house training for the public on food safety issues were also used to improve the hotel's image. One interviewee indicated that it was senior management's intention to place food safety issues in the company's strategic report, which can maintain or improve a company's image. It was also the company's policy to have a management team led by top management to run the food safety strategy more efficiently.

4. CONCLUSIONS

The main aim of this study was to understand the food safety strategies (activities) in a pioneering hotel, and the associated relationships between the motivating antecedents and food safety strategies and to determine the key motivating factors in order to provide a source of new hypotheses for further study.

This study has provided details on "what" and "how" with respect to carrying out the food safety strategies and on "how" top management commends each motivating factor. The effect of perceptions of corporate image, a firm's capability, and senior management's commitment to a food safety strategy in a large international chain hotel provided insight into the issues faced when attempting to encourage or mandate stronger safe food control systems. The results of this study suggest the most effective food safety practices and information for helping the hotel industry to further improve food safety strategies or practices. These findings can also be employed as a basis for further review by policy makers and educational institutions to ensure that the policy is appropriate for the hotel industry.

This study identified three key motivating factors— corporate image, firm's capability, and senior management's commitment—which can be developed as new hypotheses for further testing and refinement. This reduction process of data by identifying key factors is significant for forming future research.

Moreover, the result found in this study representing the relationship between the motivating factors and food safety strategies can be applied to other sectors of the food service industry or other countries confronted by similar situations and problems. The intention of this study was not to draw conclusions from the limited sample, but to provide direction for future large scale research. Therefore, any findings from the analysis of the interviewees' responses have not been used to generalise.

REFERENCES

1. World Health Organization, "International Travel and Health," 2002. http://www.who.int/ith/ITH2002.pdf

2. World Health Organization, "Food Safety," 1999.

3. C. G. Q. Chi and H. Qu, "Examining the Relationship between Tourists' Attribute Satisfaction and Overall Satisfaction," Journal of Hospitality Marketing & Management, Vol. 18, No. 1, 2009, pp. 4-25.

4. B. Ozdemir, A. Aksu, R. Ehtiyar, B. Çizel, R. B. Çizel and E. T. İçigen, "Relationships among Tourist Profile, Satisfaction and Destination Loyalty: Examining Empirical Evidences in Antalya Region of Turkey," Journal of Hospitality Marketing & Management, Vol. 21, No. 5, 2012, pp. 506-540.

5. G. Fuchs and A. Reichel, "Tourist Destination Risk Perception: The Case of Israel," Journal of Hospitality & Leisure Marketing, Vol. 14, No. 2, 2006, pp. 83-108.

6. Y. Yilmaz, Y. Yilmaz, E. T. İçigen, Y. Ekin and B. D. Utku, "Destination Image: A Comparative Study on Pre and Post Trip Image Variations," Journal of Hospitality Marketing & Management, Vol. 18, No. 5, 2009, pp. 461-479.

7. E. G. McKeown and W. B. Werner, "Content Analysis of Consumer Confidence in Food Service in Relation to Food Safety Laws,

Publicity, and Sales," Journal of Hospitality Marketing & Management, Vol. 19, No. 1, 2009, pp. 72-81.

8. M. F. Chen, "Consumer Trust in Food Safety—A Multidisciplinary Approach and Empirical Evidence from Taiwan," Risk Analysis, Vol. 6, 2008, pp. 1553-1569.

9. A. Radam, M. L. Abu and M. R. Yacub, "Consumers' Perceptions and Attitudes towards Safety Beef Consumption," The IUP Journal of Marketing Management, Vol. IX, No. 4, 2010, pp. 29-55.

10. A. Knight, M. R. Worosz and E. C. D. Todd, "Dining for Safety Consumer Perceptions of Food Safety and Eating Out," Journal of Hospitality & Tourism Research, Vol. 33, No. 4, 2009, pp. 471-486.

11. J. Sneed and D. Henroid, "HACCP Implementation in School Foodservice: Perspectives of Foodservice Directors," The Journal of Child Nutrition & Management, 2003. http://docs.schoolnutrition.org/newsroom/jcnm/03spring/sneed/

12. S. B. Banerjee, "Corporate Environmental Strategies and Actions," Management Decision, Vol. 39, No. 2, 2001, pp. 36-44.

13. S. L. Wu, "Factors Influencing the Implementation of Food Safety Control Systems in Taiwanese International Tourist Hotels," Food Control, Vol. 28, No. 2, 2012, pp. 265-272.

14. R. K. Yin, "Case Study Research: Design and Methods," 2nd Edition, Sage, Thousand Oaks, 1999.

15. P. J. Panisello and P. C. Quantict, "Technical Barriers to Hazard Analysis Critical Point," Food Control, Vol. 12, 2001, pp. 165-173.

16. L. A. Brannon, V. K. York, K. R. Robert, C. W. Shanklin and A. D. Howells, "Appreciation of Food Safety Practices Based on Level of Experience," Journal of Foodservice Business Research, Vol. 12, No. 2, 2009, pp. 1-32.

17. G. C. Ogbeide, "A Case Study of Restaurant Training Motivations and Outcomes," International Journal of Tourism and Hospitality Research, Vol. 19, No. 1, 2008, pp. 72-177.

18. J. Salazar, H. R. Ashraf, M. Tchebg and J. Antun, "Food Service Employee Satisfaction and Motivation and the Relationship with Learning Food Safety," Journal of Culinary Science & Technology, Vol. 4, No. 2, 2005, pp. 93-108.

19. A. Knight, E. C. D. Todd and M. R. Worosz, "Dining for Safety Consumer Perceptions of Food Safety and Eating Out," Journal of Hospitality & Tourism Research, Vol. 33, No. 4, 2009, pp. 471-486.

20. U. Z. A. Ungku Fatimaha, H. C. Boo, M. Sambasivan and R. Salleh, "Foodservice Hygiene Factors—The Consumer Perspective," International Journal of Hospitality Management, Vol. 30, No. 1, 2011, pp. 38-45.

21. Y. Boshy, "Action Learning Worldwide," Palgrave, New York, 2002.

22. R. Welford, "Corporate Environmental Management 1: Systems and Strategies," Earthscan, London, 1998.

8

CHAPTER

FOOD SANITATION KNOWLEDGE, ATTITUDE, AND BEHAVIOR FOR THE UNIVERSITY RESTAURANTS EMPLOYEES

Wen-Hwa Ko

Department of Restaurant, Hotel and Institutional Management, Fu-Jen University, Chinese Taipei.

ABSTRACT

The purpose of this study was to discuss the food sanitation knowledge, attitude, and behavior for the employees of university restaurants, and furthermore, to explain the interrelations of these various were occurring at school. A questionnaire survey procedure was used. Each two constructs of food sanitation attitude and behavior were analyzed by factor analysis. Data was analyzed by description, Pearson's correlation and multiple regression analysis. The correctness rate toward the whole sanitation knowledge of trials was 70%. The overall attitude toward food sanitation was prone to positive and the attitude of employee self-responsibility was superior to the attitude of food sanitation practice. The sanitary guiding behavior was better than the sanitary habit behavior; and there was a significant relationship indicated through Pearson correlation analysis among three various. The sanitation knowledge and sanitation attitude showed a 42.6% predic-tive power to behavior, the attitude was mediated between knowledge and behavior. The university restaurant employees shared a more pessimistic view toward the benefits of training and the institutions could establish a committee for monitoring food nutrition and sanitation. It provides valuable information for development employees training while seeking to raise school restaurant food safety levels.

Keywords: Food Sanitation, Knowledge, Attitude, Behavior

INTRODUCTION

School meals must provide the organism with all the necessary nutrients and be safe from the hygienic-sanitary point of view. Beside their quests for knowledge, learning and finding the right places on campus to satisfy their physical hungers is a prerequisite to college students [1]. Chen [2] studied the impact of collective behavior on college students' eat-out preferences; he found that the majority of college students had allowances to spend on food. Students were able to afford their preferential meals in restaurants based on what they liked. Huang [3] was the only one relevant study that was identified as special interest and investigated and explored the university restaurant catering employees' views held on food sanitation based on their knowledge, attitude, and behavior toward food sanitation. However, this study was lagging behind the legislative requirements announced in 2000 by the Department of Health of Executive Yuan for Good Hygienic Practice (GHP). All catering premises were governed to practice by the GHP after the year 2000. There is a great need to update academic studies to coincide to the level of the current legislative requirement imposed by the Department of Health [4]. The statistics show that 15% - 20% of food poisonings were occurring at school; the primary mode of the poisonings were caused by a lack of temperature control of food and cross contamination in Taiwan, 2008 [5]. Poor sanitary practices of food storage, handling, and preparation can create an environment in which contaminations are more easily transmitted [6]. Proper personal and food handling practices are the basis for preventing the transmission of pathogens from food handling personnel to the consumer [7]. The retail food service industry has intensified efforts to improve retail food safety through the training of restaurant employees [8,9]. The food poisoning outbreaks reveal that small and medium established sized businesses are often important locations in the transmission of food borne illnesses [10].

Schwardtz [11] suggested four types of relationships between knowledge, attitude, and behavior. The first type of relationship exists where knowledge can directly influence attitude but not directly influence behavior. The second type of relationship exists where knowledge and attitude influence each other at the same time. The third type of relationship exists where knowledge and attitude independently influence behavior. The fourth type of relationship exists where knowledge shared direct and indirect influences on behavior. Attitude was the mediating variable between knowledge and behavior. A number of studies [12,13] have indicated that although training may bring about an increased knowledge of food safety, this does not always result in a positive change in food handling behavior. It has been suggested that this disparity between knowledge and practice occurs because of much of the existing training. Rennie [14] showed that the KAP model assumes that an individual's behavior or practice (P) is dependent on their knowledge (K) and suggested that the mere provision of information will lead directly to a change in attitude (A) and consequently a change in behavior. It has been suggested that this model is flawed in its assumption that knowledge is the main precursor to behavioral change [15]. Lin and Chen [16] found when studying school children's knowledge, attitude, and behavior toward food sanitation in Taiwan that these factors interacted with each other in a positive manner. Huang [17] found that lunch-box factory employees' knowledge, attitude, and behavior toward food sanitation in Taiwan were in a positive relationship with each other. These results also found that knowledge and attitude were not positively influenced and knowledge and behavior also did not share a positive relationship. In conclusion, of the previous findings on the four types of relationships between knowledge, attitude, and behavior toward food sanitation, they were inconclusive with inconsistent findings that demonstrated the precise relationships shared between these three factors. Therefore, we will attempt to find the relationship between food sanitation knowledge, attitude, and behavior.

The purpose of this study centered on the investigation of the knowledge, attitude, and behavior of employees of university restaurants in Taipei, whereby the relationships of these factors with their determinant influences were also explored. It is hoped that such a study would adequately provide the industries and governmental agencies with blueprints for future employee training, and references for policy purposes. It shall also act as guiding indexes for selecting future university restaurants in tenders.

2. METHODOLOGY

2.1. Sampling of Participants

A quota-sampling method was adopted for data collection. The subjects were full and part time catering employees from university and college restaurants located in Taipei, Taiwan. The pre-testing of surveys was conducted in Fu Jen Catholic University's campus restaurant in November 2009. Fifty-seven questionnaires were returned. The formal run of the survey was conducted in December 2009. Two groups of random subjects were selected from 5 universities in Taipei City, and five universities in Taipei County among 45 universities from both districts. Questionnaires were given out in the designated universities' restaurants to catering employees such as food preparers, food distributors, food packaging employees, and restaurant employees. Both part-time and full-time employees were included. There were a total of 550 questionnaires given out with 493 returned, resulting in a return rate of 89.6%.

2.2. Questionnaire Development

There were four parts to the questionnaires. The first part of the questionnaire included personal background variables such as gender, age, work experience, education, certification of cooking licenses, participation in food sanitation trainings, and whether the subject was under supervision or not. The second part of the questionnaire included 20 yes/no questions on their knowledge of food sanitation. The set of questions were in reference to the knowledge base form suggested from studies conducted by Huang [3], and the exam questions for C level Chinese cuisine chef certification. With each right answer, 1 mark was allocated to the subject. No mark would be given for wrong answers. The more knowledgeable the subject was, the greater the marks. Questions were focused on two central themes that related to good standard hygiene practice and food poisoning issues. The third and forth parts of the questionnaires focused on food sanitation attitudes and behaviors in reference to the studies conducted by Lee et al. [18] and Tokuc et al. [9]. There were 15 questions in each part using a 5-point Likert Scale. The greater marks the subject rendered the higher the rank they obtained, which indicated good food sanitation behavior.

2.3. Factor Analysis, Reliability and Validity Analysis

The returned pre-test questionnaires were analyzed by factor analysis of maximum likelihood factor extraction on the attitude part. After rotation, the constructs number was 2, factor loading was greater than 0.3, and KMO value was 0.730 ($P < 0.05$). Based on these analyses, question 7 and question 8 were eliminated. The rest of the questions were divided into two constructs. The first construct consisted of item numbers 1, 2, 3, 5, 6, 10, and 14. It was called the "employee's self-responsibility attitude". The second construct consisted of item numbers 4, 9, 11, 12, 13, and 15. It was called the "food sanitation practice attitude". The cronbach's α value for "employee's self-responsibility attitude" was 0.742, and the cronbach's α value for the "food sanitation practice attitude" was 0.843. For the behavior part, the questionnaire also used the maximum likelihood factor extraction and rotation, factor loading was greater than 0.3, and KMO value was 0.697 ($P < 0.05$), therefore, question 5 was eliminated. The rest of the questions were constructed into two constructs based on variable value. The first construct consisted of questions 2, 4, 7, 8, 11, 12, 14, and 15, it was called the "food sanitation guiding behavior". The second construct consisted of numbers 1, 3, 6, 9, 10, and 13, it was called the "sanitary habit behavior". These two constructs could be interpreted as 33.63%. After conducting reliability analysis, the sanitary guiding behavior had a cronbach's α value of 0.703, whereas "sanitary guiding behavior" had a cronbach's α value of 0.747. The questionnaire surveyed on the attitude produced cronbach's α values for employee self-responsibility attitude and food sanitation practice attitude as 0.723, and 0.9. The cronbach's α value for sanitary guiding behavior and sanitary habit behavior was 0.800 and 0.734. The design of this study also utilized expert validity analysis to gain insights to contribute to the finalized formal questionnaire, which was completed in January 2008 by five experts in relative fields.

2.4. Data Analysis

This study used the SPSS 17.0 version of statistic software to conduct data analysis. The data analysis methodologies were in line with the aim and design of this study, which included descriptive covariance analysis, factor analysis, reliability analysis, correlation analysis, and multiple regression analysis.

Table 1. The socio-demographic characteristics for the restaurant employees (N = 473).

Characteristics	Description item	n	Percents (%)
Gender	Male	189	40.0
	Female	284	60.0
Age	Under 19	103	22.2
	20 - 24	86	18.5
	25 - 29	122	25.8
	30 - 34	55	11.8
	35 - 39	57	12.1
	45 - 49	36	7.6
	Above 50	13	2.7
Work experience (years)	Under 1	117	24.7
	1 - 3	213	45.0
	4 - 6	37	7.8
	7 - 9	35	7.4
	10 - 12	47	9.9
	Above 13	24	5.1
Education	Under junior high school	14	3.0
	Junior high school	36	7.6
	Senior high school	209	44.2
	College	37	7.8
	University and above	177	37.4
Cooking licenses	No	335	70.8
	Yes	138	29.2
Training class	No	370	78.2
	Yes	103	21.8
Whether or not the subject was under supervision	Yes	332	70.2
	No	141	29.8

3. RESULTS

Of the 473 respondents, most of them were female (60.0%). The age group peaked at 25 to 29 years, which accounted for 25.8%. Nearly fifty percent of the respondents had worked in the related industry for 1 to 3 years. In respect to their educational background, 44.2% were high school/technical school graduates. In respect to their participation in food sanitation trainings, 78.2% of all subjects had not had any training within a year. Most subjects had responsible employee supervision of food sanitation in their work places. Other related results are shown inTable 1.

A greater proportion of catering employees answered correctly the part that involved good hygiene practice sanitation (GHP) standard (M = 7.63).

Whereas questions involving food poisoning had lower counts of correct answers (M = 6.99). Within the expected GHP standard, 99.6% of the subjects knew that, "Disinfected water proof gloves should be worn on hands for processing uncookedready-to-eat foods." This item had the highest number of correct answers. The least correctly answered question pertained to "Freezing had a better sterilizing effect than heating." Only 14.0% of the employees answered it correctly. With respect to food poisoning, the most correct item (99.8%) was, "Lunch boxes that were not sold during lunch time can be kept until night time for reselling." The question that was answered the least correctly was "Salmonella was easily contracted from seafood products," with only a 28.8% correct rate. It was demonstrated that most restaurant employees were not familiar enough with food poisoning agents and the types of food poisoning.

There were two constructs involved in the catering employee's food sanitation attitudes, which were "employee's self-responsibility attitude" and "food sanitation practice attitude" as shown in Table 2. The average mean for "employee's self-responsibility attitude" was 4.42, and the average mean for "food sanitation practice attitude" was 4.48. This demonstrated that the majority of restaurants' catering employees had positive attitudes toward food sanitation. Among those items such as, "I think hand washing before touching food can decrease the risk of contamination," and "I think raw food and cooked food must be handled separately," had the highest scores. However, with respect to participation in food sanitation training there were different results, where answers to the questions, "To participate in food sanitation seminars is very important to me," and "I think chefs should participate in a food sanitation seminar each year," had lower scores for employee self-responsibility attitude.

Table 2. The scaling of sanitation attitudes for the employee (N = 473).

Constructs	Item	Mean	S.D.	Ranking
Employee's self-responsibility attitude	1. I think it is very important to study more food hygiene knowledge.	4.47	0.631	5
	2. It is my responsibility to offer safe hygienic foods to customers.	4.58	0.547	4
	3. I think hand washing before touching food can decrease the risk of contamination.	4.67	0.479	1
	5. I think disinfected water proof gloves could decrease food poisoning.	4.60	0.670	3
	6. Wine can get rid of the fishy smell.	4.61	0.683	2
	10. I think only the food processing handling person should be in charge of decreasing the risk of food poising.	4.27	1.016	6
	14. Participating in food sanitation training is very important to me.	3.74	1.215	7
Mean		4.42		
Food sanitation practice attitude	4. I think raw and cooked foods must be handled separately.	4.67	0.520	1
	9. I think chefs should participate in food sanitation seminars each year.	4.23	0.791	6
	11. I think reading more periodicals pertaining to hygiene are important.	4.45	0.703	4
	12. Taking hygiene classes will improve my hygiene habits.	4.43	0.641	5
	13. It is my responsibility to correctly execute GHP.	4.48	0.631	3
	15. I think it is very important for the employee to have proper hygienic habits.	4.59	0.545	2
Mean		4.48		

Table 3 shows the two constructs involved in the catering employee's food sanitation behavior, which was "sanitary guiding behavior" and "sanitary habit behavior". For "sanitary guiding behavior" the average was 4.03. For "sanitary habit behavior" the average was 4.29. Among these items in the construct of food sanitation guiding behavior, "I would do at least one health check each year," "I did not need to clean the drainage each day," "When I washed the dishes, I would use the three sinks method," and "If there were cracks on the dishes I would still use them," were not over 4 point scales (Table 3).

Table 3. The scaling of sanitation behavior for the employee (N = 473).

Constructs	Item	Mean	S.D.	Ranking
Sanitary guiding behavior	2. I will wear the tidy uniform prior to beginning work.	4.23	1.004	3
	4. I will perform at least one health check every year.	3.53	1.342	7
	7. If I have wounds on my hand, I will wear gloves.	4.55	0.738	1
	8. I will use different chopping blocks to deal with the food materials.	3.97	1.157	5
	11. I will completely disinfect the cutter and chopping block after work every day.	4.12	0.984	4
	12. I will clean and dry the facility after work everyday.	4.54	0.703	2
	14. When I wash the dishes, I will use the three sinks methods.	3.79	1.248	6
	15. I need to clean the drainage every day.	3.51	1.423	8
Mean		4.03		
Sanitary habit behavior	1. The first thing to do upon entering the kitchen is wash my hands.	4.62	0.633	1
	3. I will wash my hands when I touch the cooked food.	4.51	0.779	3
	6. I don't use cooking tools to taste the food.	4.10	1.314	5
	9. I will dispose of any moldy food.	4.57	0.878	2
	10. If there are cracks on the dishes, I will not to use them.	3.67	1.485	6
	13. I will disinfect the work area regularly.	4.24	0.906	4
Mean		4.29		

As shown in Table 4, there were positive relationships between knowledge, attitude, and behavior on food sanitation extrapolated from these university and college restaurants catering employees. Further, food sanitation attitude and food sanitation behavior were positively related. These findings correlate with the findings reported by Chen [2].

As shown in Table 5, there were significant influences between food sanitation knowledge and attitude, the entire $R^2$was 0.025, indicating that food sanitation knowledge could be interpreted into 2.5% of food sanitation attitudes. Between the influences of food sanitation knowledge and behavior, food sanitation knowledge could be interpreted into 6.4% of food sanitation behavior. However, food sanitation knowledge and attitude can be explained by 42.6% of food sanitation behavior. In this study, the independent variable (knowledge) affected the mediator (attitude) in the first equation; in the second equation, the independent variable (knowledge) affected the dependent variable (behavior) while the mediator (attitude) also affected the dependent variable (behavior). In the third equation the dependent variable regressed on both the independent variable and the mediator. According to our results, food sanitation attitude was a partial mediated effect of behavior. Compared to Table 5, it was also found that the Beta coefficient for knowledge on behavior had shown a decreased trend that indicated attitude was the partial mediator variable for knowledge on behavior. This matched the findings reported by Schwartz [11] where the fourth type of interaction, in which knowledge had a direct or indirect relationship on behavior, and attitude was the mediating variable between knowledge and behavior. It could be stipulated that knowledge, attitude, and behavior were associated with one another. Under two knowledge constructs of "good hygiene practice knowledge" and "food poisoning knowledge" together with the two constructs of "employee self-responsibility attitude" and "food sanitation practice attitude" from food sanitation attitude had significant influences on three aspects from food sanitation behavior other than the aspects of "food poisoning", "Good hygiene practice knowledge", "employee self-responsibility attitude" and "food sanitation practice attitude", which could explain 44.9% of food sanitation behavior (Table 6).

4. DISCUSSION & RECOMMENDATION

4.1. Discussion

In this study, it was found that knowledge of good hygiene practices scored better than knowledge of food poisoning. The area of food poisoning agents' transmission channels was lacking. However, with respect to participation in food sanitation training there were different results where answers to the questions. It indicates that most catering employees thought food sanitation training was not important. Special attention should be rendered to address this problem. Training catering employees elevated their sanitation knowledge. In particular, training is a good way to increase food poisoning knowledge. Food sanitation attitudes had an average score of between 3.74 and 4.67, where the attitude of the group as a whole inclined toward being more positive. Food sanitation behavior had an average score of between 3.57 and 4.57. For "sanitary guiding behavior" the average was 4.03 and "sanitary habit behavior" the average was 4.29.

From the analysis of relationships between knowledge, attitude, and behavior were extrapolated and found that they shared positive relationships with one another. The results also indicated attitude was the mediating variable for knowledge on behavior. Griffith and Clayton [19] reported that improved knowledge will lead to behavioral changes involving improved practices, and also suggested that other factors, including staff attitudes, can limit or prevent improvements in staff practices. Training for employees has been shown to improve food safety knowledge and hygienic awareness and may result in improved food safety practices [20], however, there is considerable evidence that improved knowledge does not always translate into improved food handling behavior [21]. Thus, there would appear to be a clear linkage between effective formal training, improved catering practices, and prevention or significant reduction of food borne outbreaks in the foodservice industry. Attitudes, an important factor besides knowledge and enforcement, ensure a downward trend of food borne illnesses. The necessary link of positive behavior, attitudes, and continued education of food handlers towards the sustainability of safe food handling practices has been highlighted [12]. The studies [13] have demonstrated that although training may increase food safety knowledge, a positive change does not always subsequently occur in terms of food handling behavior.

Table 4. The relationship among knowledge, attitude, and behavior for employees.

	Knowledge	Attitude	Behavior
Knowledge	1		
Attitude	0.163*	1	
Behavior	0.263*	0.636*	1

*$P < 0.05$.

Table 5. Regression analysis of food safety knowledge, attitude, and behavior for employees.

Dependent variable - Independent variable	Adjusted R^2	Standardized Beta Coefficient	t Value	Significance
Attitude - knowledge	0.025	0.163	3.592	0.000*
Behavior- knowledge	0.064	0.256	5.757	0.000*
Behavior- attitude	0.403	0.636	18.870	0.000*
Behavior	0.426			
- knowledge		0.162	4.517	0.000*
- attitude		0.609	17.011	0.000*

*$P < 0.05$.

Table 6. Regression analysis of food safety knowledge construct and attitude construct to behavior for employees.

Independent variable	Adjusted R^2	Standardized Beta Coefficient	t Value	Significance
	0.449			
Good hygiene practice knowledge		0.190	5.362	0.000*
Food poisoning knowledge		0.040	1.173	0.242
Employee self-responsibility attitude		0.127	2.450	0.015*
Food sanitation practice attitude		0.506	9.615	0.000*

F value = 98.912, $p = 0.000$, $R = .673$, $R^2 = 0.453$, Adjusted, $R^2 = 0.449$, *$P < 0.05$.

4.2. Recommendation

The university catering employees shared a more pessimistic view toward the benefits of training. It is suggested that the academic institution should take a more serious role in implementing more training programs to improve its employees' knowledge and attitude toward food sanitation. Through the cascades of intrinsic learning processes it was understood that knowledge transformed attitude where attitude further influenced behavior. Thus, nothing should be more important than to improve employees' knowledge through training where a more positive attitude can be formed that fortifies the desired food sanitation behavior. Additionally, institutions could establish a student committee for monitoring food nutrition and sanitation. Some schools have already implemented student committees for monitoring in campus restaurants. Although the responsibility of overseeing campus restaurants usually lies within the realm of the institution dietician, getting the students involved will facilitate a better emphasis and understanding for these students. Finally, institutions could increase the number of food poisoning courses, although most employees had an understanding of the legislation of good hygienic practices, but this study indicated that a lack of food poisoning knowledge was apparent. It is suggested that improvement can be seen through implementing food poisoning courses while providing sound supervision will indeed rectify this insufficiency. For the recommendation of future research, it would be helpful to conduct comparisons of the differences between on campus and off campus restaurants catering employees' knowledge, attitude, and behavior toward food sanitation. Interviews can be facilitated to probe the owner's concerns over challenges and problems faced in their business operation and how they intend to resolve these problems. It is also helpful to compare the differences between franchised chain restaurants and independent restaurants.

REFERENCES

1. A. K. C. Campos, A. M. S. Cardonha, L. B. G. Pinheiro, N. R. Ferreira, P. R. M. Azevedo and T. L. M. Stamford, "Assessment of Personal Hygiene and Practices of Food Handlers in Municipal Public Schools of Natal, Brazil," Food Control, Vol. 2 ,No. 10, 2009, pp. 807-810.

2. N. Y. Chen, "The Study of the Effects on Conformity for Purchasing Decision of Food Away-from-Home of College Students," Unpublished Master Degree Thesis, Shih Hsin University, Taipei, 2004.

3. S. Y. Huang, "A Study of F and B Employees' Food Sanitation Behavior in Different Counties," Journal of Health Education, Vol. 15, No.1, 1994, pp. 15-20.

4. Department of Health, "Good Hygiene Practice," Department of Health, Executive Yuan, ROC, Taipei, Taiwan, 2005.

5. Department of Health, "Information of Food News," 2010. http://food.doh.gov.tw/FoodNew/MenuThird.aspx?SecondMenuID=34&ThirdMenuID=298

6. J. E. Fielding, A. Aguirre and E. Palaiologos, "Effectiveness of Altered Incentives in a Food Safety Inspection Program," Prevent Medicine, Vol. 32, No. 3, 2001, pp. 239-244.

7. H. S. Evans, P. Madden, C. Doudlas, G. K. Adak, S. J. O'Brien, T. Djuretic, P. G. Wall and R. Stanwell-Smith, "General Outbreaks of Infectious Intestinal Disease in England and Wales: 1995 and 1996," Community Disease Public Health, Vol. 1, No. 2, 1998, pp. 165-171.

8. R. Lynch, B. I. Elledge, C. C. Griffith and D. T. Boatrigh, "A Comparison of Food Safety Knowledge among Restaurant Managers, by Source of Training and Experience, in Oklahoma County, Oklahoma," Journal of Environment Health, Vol. 66, No. 1, 2003, pp. 9-14.

9. B. Tokuc, G. Ekuklu, U. Berberoglu, E. Bilge and H. Dedeler, "Knowledge, Attitudes, and Self-Reported Practices of Food Service Staff Regarding Food Hygiene, in Edirne, Tyrkey," Food Control, Vol. 20, No. 8, 2009, pp. 565-568.

10. E. Walker, C. Pritchard and S. Forsythe, "Hazard Analysis Critical Control Point and Prerequisite Implementation in Small and Medium Size Food Businesses," Food Control, Vol. 14, No. 2, 2003, pp. 169-174.

11. N. E. Schwartz, "Nutrition Knowledge, Attitude, and Practices of High School Graduates," Journal of American Dietary Association, Vol. 66, No. 1, 1975, pp. 28-31.

12. M. Howes, S. McEwen, M. Griffiths and L. Harris, "Food Handler Certification by Home Study: Measuring Changes in Knowledge and Behavior," Dairy Food Environment Sanitation, Vol. 16, No. 7, 1996, pp. 737-744.

13. S. C. Powell, R. W. Attwell and S. J. Massey, "The Impact of Training on Knowledge and Standards of Food Hygiene—A Pilot Study," International Journal of Environment Health Research, Vol. 7, No. 5, 1997, pp. 329- 334.

14. D. M. Rennie, "Health Education Models and Food Hygiene Education," Journal of Royal Social Health, Vol. 115, No. 1, 1995, pp. 75-79.

15. J. E. Ehiri, G. P. Morris and J. McEwen, "Evaluation of a Food Hygiene Training Course in Scotland," Food Control, Vol. 8, No. 2, 1997, pp. 137-147.

16. J. R. Lin and S. Y. Chen, "Study of the Knowledge, Attitudes, and Practices of Food Hygiene by Cooks in Daycare Centers in the Tainan Area," Nutrition Science Journal, Vol. 29, No. 1, 2004, pp. 83-96.

17. S. Y. Huang, "A Study of F and B Employees' Food Sanitation Knowledge, Attitude and Behavior in Taiwan," Fu Jen Journal of Human Ecology, Vol. 1, No. 1, 1995, pp. 53-71.

18. S. Y. Lee, J. S. Shieh and C. A. Wen, "An Establishment for the Measurement of Food Safety Practices to Chinese Restaurant Employees," Nutrition Science Journal, Vol. 24, No. 3, 1999, pp. 288-297.

19. C. J. Griffith and D. Clayton, "Food Safety Knowledge, Attitudes and Practices of Caterers in the UK," In B. Maunsell and D. J. Bolton, Eds., Restaurant and Catering Food Safety, Teagasc, Dublin, Ireland, 2005.

20. S. Thompson, R. de Burger and O. Kadri, "The Toronto Food Inspection and Disclosure System: A Case Study". British Food Journal, Vol. 107, No. 2, 2005, pp. 140-149.

21. H. Kassa, "An Outbreak of Norwalk-Like Viral Gastroenteritis in a Frequently Penalized Food Service Operation," Journal of Environment Health, Vol. 64, No. 10, 2001, pp. 9-13.

9

CHAPTER

AN ASSESSMENT OF FOOD SAFETY NEEDS OF RESTAURANTS IN OWERRI, IMO STATE, NIGERIA

Sylvester N. Onyeneho and Craig W. Hedberg

Division of Environmental Health Sciences, University of Minnesota, 420 Delaware Street SE, MMC807, Minneapolis, MN 55455, USA

ABSTRACT

One hundred and forty five head chefs and catering managers of restaurants in Owerri, Nigeria were surveyed to establish their knowledge of food safety hazards and control measures. Face-to-face interviews were conducted and data collected on their knowledge of risk perception, food handling practices, temperature control, foodborne pathogens, and personal hygiene. Ninety-two percent reported that they clean and sanitize food equipment and contact surfaces while 37% engaged in cross-contamination practices. Forty-nine percent reported that they would allow a sick person to handle food. Only 70% reported that they always washed their hands while 6% said that they continued cooking after cracking raw eggs. All respondents said that they washed their hands after handling raw meat, chicken or fish. About 35% lacked knowledge of ideal refrigeration temperature while 6% could not adjust refrigerator temperature. Only 40%, 28%, and 21% had knowledge of *Salmonella*, *E. coli*, and Hepatitis A, respectively while 8% and 3% had knowledge of *Listeria* and *Vibrio* respectively, as pathogens. Open markets and private bore holes supplied most of their foods and water, respectively. Pearson's Correlation Coefficient analysis revealed almost perfect linear relationship between education and knowledge of pathogens (r = 0.999), cooking school attendance and food safety knowledge (r = 0.992), and class of restaurant and food safety knowledge (r = 0.878). The lack of current knowledge of food safety among restaurant staff highlights increased risk associated with fast foods and restaurants in Owerri.

Keywords: Restaurants; Food safety; Foodborne pathogens; Hygiene; Cross-contamination

1. INTRODUCTION

Although the safety of foods served in restaurants in Nigeria has been an ongoing concern, the unhygienic environments where some restaurants are located contribute to transmission of foodborne diseases. Health experts in Nigeria have said that the shortage of water may affect the success of any safety campaigns. According to the WHO/UNICEF Joint Monitoring Program (JMP) for water and sanitation [1], only 58 percent of the country's 160 million people have access to potable water. Food safety issues in Africa are mostly centered on illnesses that are linked to poor hygiene but food hygiene in homes, schools and markets remains an area of concern. Although outbreaks are frequent in the African region, individual countries have done little to implement surveillance systems for food-borne diseases [2]. Some of the challenges to any surveillance system in the country include the fact that in many parts of Nigeria, food poisoning is usually associated with evil spirit, malice or curses [3]. Also, the culture of self-prescription prevents cases from being treated at hospitals. Symptoms of foodborne illnesses such as vomiting, diarrhea, fever are regarded as "common diseases" and everyone seems to know the cure and requires only going to the "chemist" to purchase self-prescribed medication. There is lack of proper monitoring and supervision by food safety officers and the enforcement of food hygiene regulations. In recognition of the importance of food safety as an essential factor for achieving a high level of health for all Nigerians, the Government of Nigeria launched the National Policy of Food Hygiene and

Safety in 2000 as an integral part of the Nigerian National Health Policy. The overall goal of this policy was the attainment of a high level of food hygiene and safety practices that will promote health, control food-borne diseases, minimize and finally eliminate the risk of diseases related to poor food hygiene and safety [2]. In most developing countries including Nigeria, urban adults eat food from street vendors regularly because it is easily available, affordable and usually fresh. Street-vended foods may pose significant public health problems due to lack of basic infrastructure and services, such as potable water supplies, their temporary nature, and poor knowledge of basic food safety measures [4]. Many of the food vendors are itinerant in nature and vary from mobile carts to fixed stalls and food centers. The operators push their food-laden carts, wheel-barrows or specially-designed bicycles from one location to another often "parking" under tree shades to serve their customers. Food vending and food hawking have been an integral and important part of the business culture of Igbo people who inhabit Imo State. They provide a source of inexpensive, convenient and often nutritious food for urban and rural poor. They are also a source of attractive and varied food for tourists and the economically advantaged. Most importantly, they are a major source of income for a vast number of persons, a chance for self-employment, and the opportunity to develop business skills with low capital investment. Besides offering business opportunities for developing entrepreneurs, the sale of street foods can make a sizeable contribution to the economies of developing countries [5].

One obstacle to food safety in Nigeria is refuse disposal and lack of toilet facilities for the customers. Most of the eating stalls around markets in Nigeria are characterized by unsanitary conditions, including poor water supply and poor drainage systems, unsanitary waste disposal and overcrowding, resulting in poor personal and environmental hygiene [6]. Another area of food safety concern is the source foods, and ingredients supply. Most of these establishments purchase their raw materials and ingredients in the open markets. Foods, especially meat, fish and ingredients are often displayed openly on tables and on the ground in very poor sanitary environments, while unwashed fingers are used to feel foodstuffs and ingredients for texture and to ascertain the adequacy of manual grinding. The prevalence of flies at the markets and the apparent lack of facilities for food protection suggest a high potential for contamination in preparation, facilities and infrastructure [7].

Although Nigeria has no official foodborne disease surveillance system, cases of food poisoning resulting in deaths and hospitalizations have been reported. An outbreak of food poisoning in Ibadan, Nigeria, claimed about 20 lives and a new phage type U282 of *Salmonella typhimurium*, the causative organism, was isolated from a sandwich filling [8]. Cases of food poisoning among three families in Kano due to yam flour consumption were reported and investigations indicated that the use of certain lethal preservatives for the processing of the yam flour might be responsible [9]. Another food poisoning report attributed to yam flour consumption in five families in Ilorin, central Nigeria was also reported [10]. The rising cases of food poisoning in Nigeria were linked to the misuse and abuse of agrochemicals and pesticides on grains and other agricultural products. The report stated that these chemicals were wrongly applied and abused in preventing pests. Over three million cases of acute food poisoning and twenty-thousand deaths occur annually due to exposure of food pesticides. An acute onset of gastrointestinal symptoms among people who had attended and eaten at a burial ceremony resulted in 60 case patients and 3 deaths [11]. The WHO estimates that more than 200,000 people die of food poisoning annually in Nigeria from foodborne pathogens (especially *E. coli* and *Salmonella*). The deaths were caused by contaminated foods through improper processing, preservation and service [12].

Owerri is the capital city of Imo State in southeastern Nigeria. Owerri had many registered hotels and school cafeterias but, hundreds of eating houses, food canteens, food kiosks and other food vending stalls were scattered in and around the city. Most of the hotels operated restaurants while most of the schools had cafeterias. However, street kiosks with poor hygiene and food handling practices are often the sources of lunch for school children in Nigeria who attend schools that do not operate school-based cafeterias. Despite these benefits, food safety is still a major concern with street foods. Most of these foods are generally prepared and sold under unhygienic conditions, with limited access to safe water, sanitary services, or garbage disposal facilities. Owerri lies within an area with a modified rainforest climate and this temperature-humidity combination favors year-round luxuriant microbial growth. Food pathogens of concern to public health include pathogenic strains of bacteria, viruses, helminthes, protozoa and algae, and certain toxic products they may produce. Foodborne infections are caused when microorganisms are ingested, and these can multiply in the human body. Intoxications result when microbial or naturally occurring toxins are consumed in contaminated foods [13]. Because restaurants with food safety-certified kitchen managers are less likely to be associated with foodborne illness outbreaks than restaurants without these managers and, because restaurants with food safety-certified kitchen managers are less likely to have critical violations on their food safety inspections than restaurants without these managers [14], this investigation was undertaken with the following objectives in mind:

(i) To ascertain the areas where food safety knowledge is lacking among restaurant managers and head cooks, and (ii) To identify the main areas in these eating places where hygiene and food safety need improvement.

2. MATERIALS AND METHODS

2.1. Subjects

Managers and head chefs responsible for food hygiene of four classes of restaurants in the city of Owerri, the capital of Imo state, Nigeria were surveyed to establish their knowledge of food safety protocols. Face-to-face interviews were conducted to collect data on their knowledge of five elements of food safety: risk perception, food handling practices, temperature control, foodborne disease pathogens, and personal hygiene. The sources of their food and water supplies were also determined. Managers of four classes of restaurants were surveyed for this study: Class A—restaurants in major hotels; Class B—school cafeterias; Class C—regular/fast food type; and Class D—"bukas" or "bukaterias" (food kiosks, roadside food sellers and itinerant food sellers or food hawkers). Clearance to use human subjects for this study (head chefs locally known as "head cooks and/or hotel managers" as well as cafeteria managers of some schools in Owerri was obtained from the Institutional Review Board (IRB) of the University of Minnesota. Restaurants in government-registered hotels, school cafeterias, fast food-type eating houses, food canteens, food kiosks and other food vending stalls in Owerri were included in the study.

2.2. Survey Administration

For this survey, the city of Owerri was divided into six zones: central, north, south, east, west, and elementary/secondary schools. Face-to-face interviews of managers of restaurants and school cafeterias in each zone were conducted to obtain data on training, food handling, personal hygiene, temperature control, food preparation, the concept and application of HACCP, and knowledge of relevant bacterial pathogens. Most of the thirty-three survey questions were single-select, multiple choice format. The questions were selected from the 2006 U.S. FDA/FSIS Food Safety Survey Topline Frequency Report [15]. Some of the questions were modified to reflect the local terminologies, foods, and ingredients normally used for cooking in Nigeria. The questionnaire was divided into the following six sections: A: Risk Perception; B: Food Handling Practices; C: Temperature Control; D: Microorganisms, Pests and Supplies; E: Personnel Hygiene and Facilities and F: Demographics. Data was collected only from consenting restaurants that did not require any pre-conditions (Table 1).

Table 1. Demographic Data.

Restaurant Class	Surveys Completed	Males (%)	Females (%)	College Ed. (%)	Cooking Sch. Ed. (%)	Average Restaurant Experience	Average Employees	Average Weekly Customers
A (Hotel-based)	40	75	25	100	100	10	20	1,500
B (School Cafe)	30	0	100	100	100	15	25	2,500
C (Fast-food)	50	29	71	30	0	25	10	4,000
D (Bukaterias)	25	0	100	0	0	8	3	600
Total	145							

3. RESULTS

Two hundred survey questionnaires and consent forms were distributed to the managers but one hundred and forty-five consented questionnaires (73%) were returned (Table 1). None of the respondents provided his or her age but all marked "adult" as a response. In class A (restaurants in major hotels), 75% of the managers were males and 25% were females. All were educated (to a minimum technical college level), all went to cooking school, and had completed a Food Hygiene course. The average restaurant experience of managers was 10 years. The restaurants had an average of 20 employees and served an average of 1,500 customers per week. In class B restaurants (school cafeterias), all the managers were female, educated and had cooking school and Food Hygiene training credentials. Managers averaged 15 years restaurant experience. Schools had an average of 25 employees and served an average of 2,500 school children every week. The owners/mangers of class C (fast-food type restaurants) were 10 males and 25 females. Fifty-seven percent had high school education and 30% indicated they had technical college education. No manager in this category attended Cooking School or Food Hygiene training. Their average restaurant experience was 25 years. These restaurants had an average of 10 employees and served an average of 4,000 customers per week. All the managers/owners of class D restaurants or "bukaterias", were females, and had only primary school education. None of them attended Cooking School or Food Hygiene training. They averaged 8 years of restaurant experience. They had an average of three employees and served an average of 600 customers per week.

3.1. Risk Perception

Over 72% of the respondents reported that it was very common for people in Nigeria to get foodborne illnesses because of the way food is prepared in their homes while about 16% said that it was somewhat common. In contrast, 33% of the respondents reported that it was very common to get foodborne illness from restaurant food. Most of the respondents agreed on the circumstances that may lead to food-borne illnesses: not cleaning and sanitizing food equipment, utensils and food contact surfaces (92%), not washing hands before cooking or serving food (88%), eating undercooked or raw food (80%). Only 61% of the respondents believed that cutting raw meat, cooked meat and fresh fruits and vegetables on the same cutting board surface could cause food-borne illness, and only 51% of the managers indicated that allowing a sick person to cook food could lead to food-borne illness (Table 2).

Table 2. Circumstances that may lead to food-borne illness (% Responses).

Food Risk	Yes	No	Don't Know
1. Eating undercooked or raw meat	80	11	9
2. Not washing hands before handling food	88	8	4
3. Cutting meat, fish, and vegetables on the same cutting board surface	61	39	0
4. The way food is cooked at home	88	8	4
5. Eating restaurant food	63	31	6
6. Not cleaning and sanitizing Equipment and utensils	92	8	0
7. Allowing a sick person to cook food	51	47	2

3.2. Food Handling Practices

Seventy percent of survey respondents reported that they always washed hands with soap before preparing foods, 20% most of the time and 10% some of the time (Figure 1). About 6% of the respondents (all in class D) would continue cooking without washing, rinsing or wiping their hands after they have cracked raw eggs (Figure 2). Eighty-two percent reported that they usually rinsed or wiped their hands after cracking raw eggs. However, only 4% of the respondents (school cafeteria only) reported that they washed their hands with soap and water after cracking raw eggs before they resumed cooking. None of the respondents reported that they continued cooking after handling raw meat, raw chicken, or raw fish without first cleaning their hands. Forty-five percent (classes C and D only) reported that they rinsed or wiped their hands clean while 55% said that they washed their hands with soap before they resumed cooking. Managers that reported washing hands with soap and water before they resumed cooking by class of restaurants were 100%, 100%, 20%, and 17% for classes A, B, C, and D, respectively).

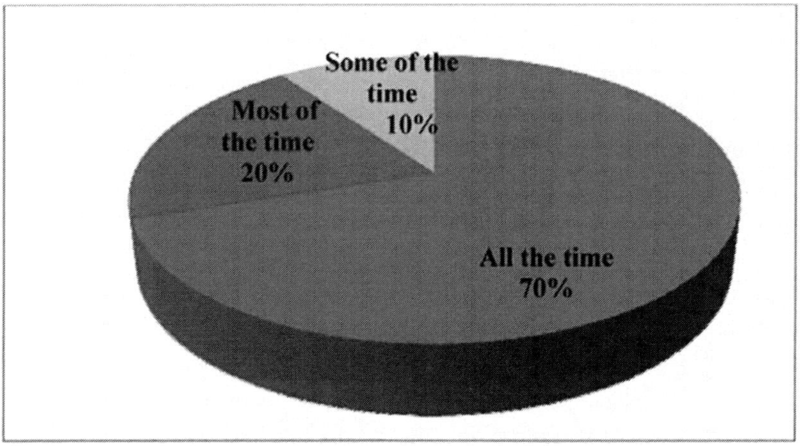

Figure 1. Frequency of handwashing before preparing food.

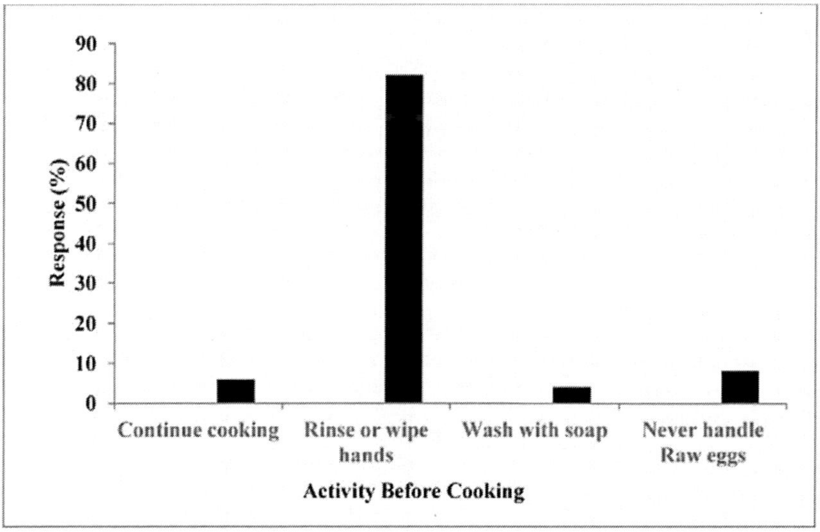

Figure 2. Sanitation activity after cracking raw eggs.

Regarding all the fruits and vegetables purchased for their restaurants, 20% of the respondents believed that all of them had been washed prior to purchase, 50% thought that most of them had been washed while 30% thought that only some of them had been washed prior to purchase. There was a wide array of methods used by the respondents to wash or rinse fruits and vegetables before using them to cook, use them for salad or serve them to customers. The methods reported included rubbing them under running water (32%), holding them under running water (3%) and, soaking them in water containing vinegar or salt (20%). After using a cutting board or other surfaces to cut raw meat, raw chicken, or raw fish, 35% reported that they rinsed the board with warm water before using the same board to cut tomato, or vegetables for salad. However, 65% reported that they washed and sanitized the board before using the same board to cut vegetables.

3.3. Temperature Control

Because all potentially hazardous foods must be kept at an internal temperature below 45 °F (7.2 °C) or an internal temperature above 140 °F (60 °C) during display and service, it was necessary to determine if the managers of these eating places possessed the knowledge of temperature control. The United States Food and Drug Administration stipulated that the maximum temperature inside a refrigerator could be set at 41 °F or 5 °C. Most of the restaurants (90%) had at least one refrigerator, a stove and/or microwave oven at their premises. Those that did not have a refrigerator were all in Class D. Refrigerators were used to store perishable foods on a short-term basis. The main purpose of refrigeration was to slow down the growth of bacteria. The results revealed that about 38% of the restaurant managers had accurate knowledge of ideal refrigerator temperature while 36% did not know anything about refrigeration temperature (Figure 3).

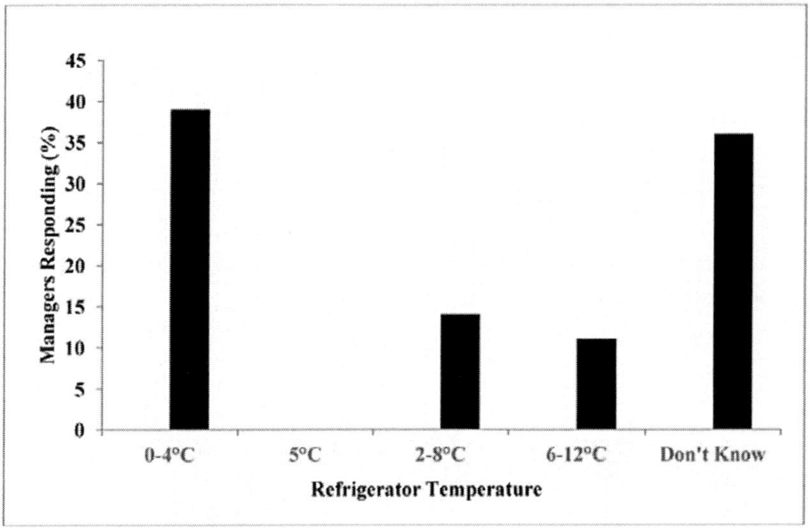

Figure 3. Knowledge of Ideal Refrigerator Temperature.

After cooking food with chicken, meat or fish that would need to be saved for the next day, 15% of the respondents reported that they put it in the refrigerator immediately after cooking while 65% left the food to cool at room temperature before storing them in the refrigerator. Of the later, 55% left the food to cool at room temperature for less than two hours while 35% left it to cool for two hours or more before storing it in the refrigerator. Leftover foods were handled in different ways by the restaurants. Sixty-four percent of the respondents reported that leftover food was cooled and stored in the refrigerator, 8% allowed the food to cool and froze it while 28% discarded any food that was cooked and not eaten. None of the restaurants served refrigerated foods cold. Thirty-nine percent warmed the food below 63 °C, 22% heated the food above 63 °C while 39% did not know the proper way to reheat the food before serving.

3.4. Knowledge of Foodborne Pathogens

About 40%, 28% and 21% of the managers indicated that they had any knowledge of *Salmonella*, *E. coli,* and Hepatitis A, respectively as agents of food-borne illnesses (Figure 4). Only 8% and 3% had any knowledge of*Listeria* and *Vibrio*, respectively, as food-borne pathogens. If food contained *Salmonella* or any of the listed pathogens, 45% of the respondents reported that the food could be made safe by cooking while 22% said that the food could be made safe by adding vinegar or lemon juice to the food. Of the managers surveyed, 25% do not know how a contaminated food could be made safe while 8% reported that the food could not be made safe for eating.

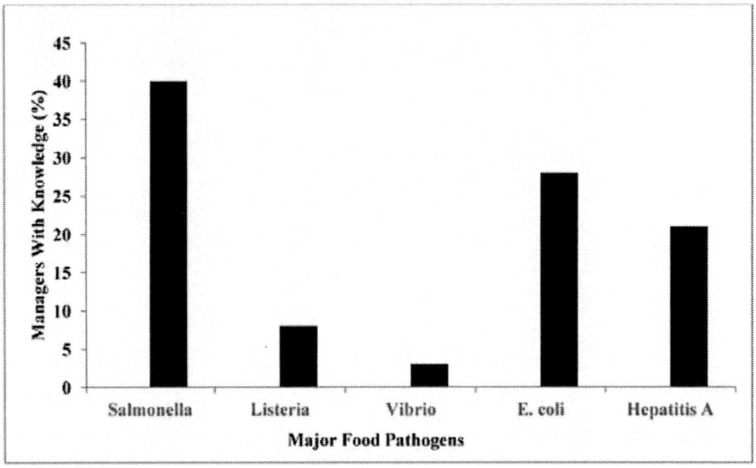

Figure 4. Managers' knowledge of major food pathogens.

3.5. Personal Hygiene and Sanitary Facilities/Equipment

Respondents almost unanimously (97%) agreed that their catering staff always paid close attention to personal hygiene. Only 72% of the restaurants surveyed had working toilet facilities on site. There were no toilet facilities for the convenience of both the employees and customers in 28% of the restaurants. Of the 28% restaurants without toilets, the majority were in class D (90%) while 10% were in class C. Most of the restaurants (92%) had hand washing facilities. Those that did not were all in class D.

3.6. Water and Materials Supply

Fifty-nine percent of the restaurants got their water from private bore holes while 41% depended on the city water supply for their water needs. The quality of water from the private bore holes had not been determined. It was also not known if this water received any treatments prior to use. Some of the restaurants also used water from the wells to prepare their foods.

The majority of the restaurants (59%) purchased their foods, ingredients and supplies daily from the open markets, 22% used regular suppliers, 14% used major registered contractors, while 5% (mostly school cafeterias) used licensed food distributors as suppliers. Flies (67%) and cockroaches (53%) were the top pests that the restaurants always encountered in their premises.

4. DISCUSSION

This study evaluated the knowledge base for prevention of food-borne illnesses at foodservice operations in Owerri because disease prevention was a crucial function of food service personnel. Our foodservice survey results indicated a general lack of knowledge of the pathogens and food safety practices that were usually associated with foodborne disease outbreaks especially in tropical environment (Figure 4). In particular, these deficiencies appeared to be greatest for restaurants most likely to be used by local residents, fast food restaurants and bukaterias.

The results from the Owerri restaurant managers' survey provide a snapshot of food safety knowledge and practice in Nigeria. From all perspectives, the results indicated that there was great need for education of foodservice personnel in Nigeria on how to prevent foodborne illnesses. The survey found that most respondents knew the circumstances that may lead to foodborne illness and the importance of hand washing. About two-thirds of the respondents understood the dangers of cross-contamination but almost half of the managers indicated that allowing a sick person to prepare or serve food could not cause people to get foodborne illness. That appeared to be one of the danger zones.

The survey found that hand washing was not practiced by the restaurant employees all the time. Only about 70% actually washed their hands all the time (Figure 1). The negative consequences of not washing hands with soap all the time prior to cooking or serving foods to customers could not be underestimated. Although 82% of the respondents indicate that they normally rinsed or washed their hands after cracking open raw eggs, only 40% and 60% of class C and D restaurants, respectively washed hands after handling open raw eggs before they resumed cooking or served foods to customers (Figure 2). Interestingly enough, none of the restaurant employees continued to cook after handling raw meat, raw chicken or raw fish. All the restaurants in classes A and B washed their hands with soap after touching raw meat, chicken and fish. Again, in classes C and D restaurants, only 20 and 17 percent, respectively, adhered to the correct hand washing protocols. One skill necessary to avoid cross contamination was to avoid using the same cutting board surface to cut raw meat, chicken or fish and fruits and vegetables. Sixty-five percent of the restaurants washed and sanitized the cutting boards each time they used them to cut raw meat, chicken or fish while 35% simply rinsed the cutting boards with warm water and used them to cut fruits and vegetables. Because cutting boards, especially those made of wood, can very easily become a source of contamination, rinsing alone even with hot water, was not sufficient to remove all the pathogens that hid in many of the tiny holes and creases of the board. This made it imperative to wash and sanitize all cutting boards after each use. This was a rule that managers must enforce in all the restaurants. Although 97% of the managers stated that their catering staff always paid close attention to personal hygiene, and 92% had hand washing facilities, this rule might be difficult to enforce in class D restaurants where 90% reported that they had no toilet facilities onsite, perhaps due to their non-permanent locations. These eating huts locally called "bukas" were usually staffed by two or three people who lack the education and training necessary to ensure food safety. Most of this type of restaurants could be termed "itinerant" because they usually move from one location to another where there are hungry people who could not afford to eat in standard and expensive restaurants. Some have permanents huts located in such places as motor parks, taxi stations, bus stations, construction sites and fields where sports activities such as soccer matches took place. Most were on wheeled carriages that could be pushed by hand. Recently, some bukas operated from a van or truck. Foods were usually cooked at home but are kept warm in insulated containers. They also have portable gas cookers for onsite cooking and warming of food. It might be difficult for this type of foodservice facilities to follow sanitation rules. In this regard, sanitation efforts could be greatly facilitated when the owners/managers were compelled to attend a crash course in sanitation and food safety as a pre-requisite for operation of their business. Also, frequent inspections (unannounced or scheduled) by government officials would make the owners put greater efforts on sanitation.

4.1. Education and Food Safety Knowledge

Based on the results of this survey, the proximate reason for the great variations in contemporary food safety knowledge and application of sanitary and hygienic standards among the restaurant managers in Owerri appeared to be the effect of training and education (Figure 5). Pearson's Correlation Coefficient ($r = 0.999$) indicated that education was directly related to knowledge of the pathogens that caused food-borne illnesses. Training obtained by attending a Cooking School also directly affected the contemporary food safety knowledge of the restaurant managers. Managers who attended cooking school appeared to have greater knowledge of food safety than those who did not. Pearson's Correlation Coefficient ($r = 0.9927$) indicates a linear relationship between Cooking School training and food safety knowledge. The survey also showed that managers with cooking school credentials had better food safety knowledge than those who did not attend cooking school. A plot of the responses of managers in different classes of restaurants and food safety knowledge is shown in Figure 5. Again, Pearson's Correlation Coefficient ($r = 8{,}787$) indicates a very strong linear negative relationship between class of restaurants and food safety knowledge.

The results of this survey also supported the study that showed that food stall personnel had four times significantly higher odds of having poor knowledge of food safety [16]. The main reason of this was that food handlers operating food stalls or food hawking activities had low level of education and were not trained in cooking schools. Again, this assessment revealed deficiencies in attitudes, knowledge and practices in the areas of cooling/reheating, temperature control and cross contamination among the restaurant managers. It was determined that although many people did not know the basic rules of food hygiene, survey respondents had knowledge of which foods were at high risk from food poisoning, but lacked the knowledge about how a food could be made safe to eat [17].

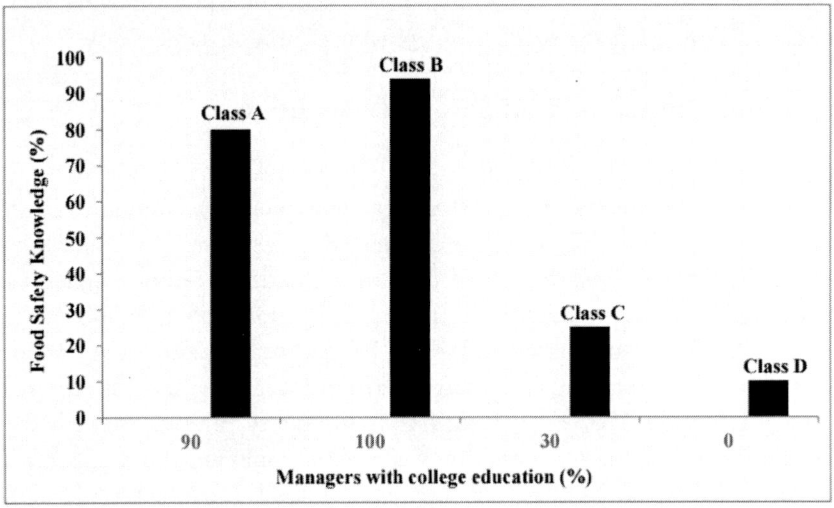

Figure 5. Relationship between managers's education, restaurant class and food safety knowledge.

4.2. Food Safety Implications

The results of this investigation presented several issues related to food safety in restaurants in Owerri. It might be difficult to ignore the fact that a significant number of foodservice workers in Owerri restaurants lacked both the education and training necessary to apply measures that would ensure the safety of foods served to their customers. In the last decades, the epidemiology of foodborne diseases changed with new or unexpected pathogens often emerging on a countrywide or worldwide scale. New foods expanded the range of potential vehicles of pathogens, wider social contexts became involved and new classes of individual became at risk

[13]. This study provided some insights as to the areas of food safety that might need intensive focus during any possible training or workshops designed to educate these restaurant workers. Experts had noted that providing tailored scientifically sound and updated knowledge and identifying factors that could contribute to the generation of positive attitudes and motivate behavior change in a definite setting could help to minimize foodborne hazard and enhance the practical utility of hygiene and training for the personnel involved in foodservice functions. Several food safety issues of concern were noteworthy: hand washing, cross-contamination, temperature control, basic knowledge of pathogens of importance in food safety, personal hygiene and general food handling practices. Food safety education was an important part of extension programming because it targeted the education of consumers and food service personnel as a strategy for preventing food-borne illnesses.

4.3. Food and Ingredients Supply

One other area of food safety concern is how these food vendors, canteens and restaurants purchased foods, and ingredients. The survey results highlighted potential dangers of the supply sources of food and ingredients. Most of the restaurants reported that they purchased their foods and ingredients in open markets. Only school cafeterias and restaurants in major hotels had licensed suppliers. Some restaurants indicated that they depended on contractors to supply their foods and ingredients. These contractors went to the open markets on a daily basis to purchase the foods which were usually fresh, and then supplied them to the restaurants. The sanitary conditions of the foods could not be guaranteed and the open markets in Nigeria were exactly what they were called. They were located in open spaces with little or no plumbing or electricity. Butchers displayed their meats without covering them on tables while meat sellers and customers constantly touched the meat with bare hands and flies were a common sight. The hot and humid tropical environment of Nigeria was ideal for bacterial and fungal growth. It was also reported that foods and ingredients were often displayed openly on tables and on the ground in very poor sanitary environments [7]. The prevalence of flies at the markets and the apparent lack of facilities for food protection suggested a high potential for contamination. These authors also reported that unwashed fingers were normally used to feel foodstuffs and ingredients for texture and to ascertain the adequacy of manual grinding. This practice also had the potential to contribute to the microbial load of the food.

4.4. Education and Training

The survey results indicated that most of the restaurant managers, especially in classes C and D, did not receive the appropriate education and training on food safety and one of the factors impacting food workers' and managers' safe food preparation practices were food safety education and training. It had been established that workers in restaurants that provide food safety training were more likely to wash their hands when they should than workers in restaurants that did not provide this training [18]; restaurants with food safety–certified kitchen managers were less likely to be associated with foodborne illness outbreaks than restaurants without these managers and that restaurants with food safety–certified kitchen managers are less likely to have critical violations on their food safety inspections than restaurants without these managers [19]. One other convenient approach to the training of employees in food safety protocols is the "Train the Trainer" model. In this model, the manager, head chef or an educated employee undergoes the training program and is given training materials. Upon return, he/she conducts in-house training for other employees. Post-training assessments demonstrated that attendees (restaurant supervisors) increased their understanding and commitment to health and safety, and felt prepared to provide health and safety training to their employees. A study of food safety practices in assisted-living facilities in Iowa concluded that training programs, both basic and food safety and the HACCP supported improvement of safe food-handling practices and implementation of prerequisite and the HACCP programs [20]. In addition, it is important that head chefs/catering managers and other personnel in key positions to deliver essential standards in consumer food safety be supported through additional training and routine inspection to ensure that appropriate knowledge is acquired and effectively applied [21]. However, numerous studies document that education alone may not result in behavioral changes, and that to change most complex behaviors, multifaceted approaches were needed. Food safety education is most effective when messages are targeted toward changing behaviors most likely to result in foodborne illness. Some Nigerian food experts have called for continuous sensitization and training of food handlers on how to operate in hygienic environment. Their approach was to sensitize people about the enormity of the problem and then train them on how to handle food properly and safely [22]. Effective restaurant managers must supervise their staff to comply with safety and sanitary protocols involving the major control factors for pathogens. These include personal hygiene and preventing sick people from handling foods, adequate cooking (time and temperature), avoiding cross-contamination including having separate utensils to prepare foods for customers with food allergies, keeping food at safe temperatures and in clean containers, and avoiding foods from unsafe sources such as open markets. Foods, water and supplies must be obtained from sources with proven safety records. Greater manager accountability for employee noncompliance and an increased emphasis on employee education could help restaurants minimize threats to public health and strengthen the restaurant industry [23].

5. CONCLUSIONS

Results of this survey highlighted key prevention and safety issues that increased the likelihood of outbreaks of food-borne diseases originating from many of these restaurants. To help mitigate these risks, restaurants should be targeted for intensive inspection and improvement of the sanitary conditions of their facilities. Furthermore, the survey highlighted the potential risks of the sources of food and water supply. The lack of current knowledge of food safety among the restaurant staff in Owerri called for the development of an appropriate training program for food service personnel in order to avert the potential danger of food-borne illness outbreaks. Government involvement might include public service announcements promoting food safety and offers of some types of incentives to encourage restaurants to participate in trainings, especially fast-food restaurants and itinerant food hawking operators.

REFERENCES

1. WHO/UNICEF Joint Monitoring Programme (JMP) for Water and Sanitation. Available online: http://www.unicef.org/ wash/index_ documents. html (accessed on 17 May 2013).

2. Omotayo, R.K.; Denloye, S.A. *The Nigerian Experience on Food Safety Regulations*; FAO/WHO Global Forum of Food Safety Regulations: Marakesh, Morocco, 2002; pp. 28–30. Available online: http://www.fao.org/ docrep/ meeting/004/ ab538e.htm (accessed on 15 May 15 2013).

3. Oyemade, A.; Omokhodion, F.O.; Olawuyi, J.F.; Sridhar, M.K.C.; Olaesha, I.O. Environmental and personal hygiene practices: Risk factors for diarrhea among children of Nigerian market women. *J. Diarrheal Dis. Res.* 1998, *16*, 241–247.

4. Sneed, J.; Strohbehn, C.; Gilmore, S.A. Food Safety Practices and readiness to implement HACCP Programs in assisted-living facilities in Iowa. *J. Am. Diet Assoc.* 2004, *104*, 1678–1683.

5. Rheinländer, T.; Olsen, M.; Bakang, J.A.; Takyi, H.; Konradsen, F.; Samuelsen, H. Keeping up appearances: Perceptions of street food safety in urban Kumasi, Ghana. *J. Urban Health* 2008, *85*, 952–964.

6. Raab, C.A.; Woodburn, M.J. Changing risk perceptions and food handling practices of Oregon household food preparers. *J. Consum. Stud. Home Econ.* 1997, *21*, 117–130.

7. Ehiri, J.E.; Azubuike, M.C.; Ubaonu, C.N.; Anyanwu, E.C.; Ibe, K.M.; Ogbonna, M.O. Critical Control Points of complementary food preparation and handling in eastern Nigeria. *Bull. World Health Organ.* 2001, *75*, 423–433.

8. Osagbemi, G.; Abdullahi, A.; Aderibigbe, S. Knowledge, attitude and practice concerning food poisoning among residents of Okene Metropolis, Nigeria. *Res. J. Soc. Sci.* 2010, *1*, 61–64.

9. Adedoyin, O.T.; Ojuawo, A.; Adesiyun, O.O.; Mark, F.; Anigilaje, E.A. Poisoning due to yam flour consumption in five families in Ilorin, central Nigeria. *West Afr. Med. J.* 2008, *27*, 41–43.

10. Adeleke, S.I. Food poisoning due to yam flour consumption in Kano (Northwest). *Nigeria Online J. Health Allied Scs.* 2009, *8*, 10.

11. Fatiregun, A.A.; Oyebade, O.A.; Oladokun, L. Investigation of an outbreak of food poisoning in a resource-limited setting. *Trop. J. Health Sci.* 2010, *17*, 1117–4153.

12. World Health Organization (WHO). *Global Burden of Disease*; WHO: Geneva, Switzerland, 2009. Available online: http://www.who.int/ healthinfo/global_burden_disease/GBD_report_ 2004update_ full.pdf (accessed on 17 May 2013).

13. Bamaiyi, P.H. Emerging food pathogens: A case study of *E. coli* 0104:H4. *Cont. J. Anim. Vet.Res.* 2011, *3*, 21–27.

14. Green, L.B. EHS-Net restaurant food safety studies: What have we learned? *J. Environ. Health* 2013, *75*, 44–45.

15. Lando, A.; Verrill, L. 2006 FDA/FSIS Food Safety Survey Topline Frequency Report. Available online: http://www.fda.gov/ Food/FoodScienceResearch/ ConsumerBehaviorResearch/ucm 080374.htm (accessed on 9 May 2011).

16. Zain, M.M.; Naing, N.N. Socio demographic characteristics of food handlers and their knowledge, attitude and practice towards food sanitation: A preliminary report. *Southeast Asian J. Trop. Med. Public Health* 2002,*33*, 410–417.

17. Raab, C.A.; Woodburn, M.J. Changing risk perceptions and food handling practices of Oregon household food preparers. *J. Consum. Stud. Home Econ.* 1997, *21*, 117–130.

18. Cates, S.C.; Muth, M.K.; Karns, S.A.; Penne, M.A.; Stone, C.N.; Harrison, J.E.; Radke, V. Certified kitchen managers: Do they improve restaurant inspection outcomes? *J. Food Prot.* 2009, *72*, 384–391.

19. Bush, D.; Paleo, L.; Baker, R.; Dewey, R.; Toktogonova, N.; Cornelio, D. Restaurant supervisor safety training: Evaluating a small business training intervention. *Public Health Rep.* 2009, *124*, 152–159.

20. Sneed, J.; Strohbehn, C.; Gilmore, S.A. Food Safety Practices and readiness to implement HACCP Programs in assisted-living facilities in Iowa. *J. Am. Diet Assoc.* 2004, *104*, 1678–1683.

21. Hedberg, C.W.; Smith, S.J.; Kirkland, E.; Radke, V.; Jones, T.F.; Selman, C.A. the EHS-Net Working Group. Systematic environmental evaluations to identify food safety differences between outbreak and nonoutbreak restaurants. *J. Food Prot.* 2006, *69*, 2697–2702.

22. AMAC. Sensitization program for food vendors operating in the Abuja Municipal Area Council (AMAC).*Premium Times* 2012, 1.

23. Dundes, L.; Swann, T. Food safety in fast-food restaurants. *J. Hum. Resour. Hosp. Tour.* 2008, *7*, 153–161.

10
CHAPTER

CUSTOMER PERCEPTION ON NUTRITIONAL INFORMATION IN RESTAURANT MENU

Norrina Din[a], Mohd Salehuddin Mohd Zahari[b], Shazali Mohd Shariff[b]

[a] *Faculty of Hotel Management and Tourism, Universiti Teknologi MARA, Permatang Pauh, Penang, Malaysia*

[b] *Faculty of Hotel Management and Tourism, Universiti Teknologi MARA, Shah Alam, Malaysia*

ABSTRACT

As more customers are educated therefore they are picky about daily food intake. The customer are demanding for more information to be displayed not only for the package food and to date they are looking at nutritional information to be included in the menu. Consequently, this study sought to investigate customer perception towards providing the nutritional information in full service restaurants menu and how it contributes to their purchase decisions. These findings produced significant indicators not only to the existing restaurant operators but also to those individuals or companies who intent to venture into foodservice businesses.

Keywords: Nutritional menu; Full service restaurant; Restaurant customer; Customer perception

1. INTRODUCTION

Most nutritionists expressed too much, imbalance or substantive nutrition intake not only lead to obesity but also contribute to chronic diseases and lessen the longevity (Gopalan, 1992; MOH, 2006). Large numbers of the world populations are reported of suffering heart diseases, hypertension, and diabetes because of obesity (Schofield & Mullainathan, 2008) and six of the ten leading causes of death are causes by nutrition related components intakes (Dennison, Dennison & Frank, 1994). Owing to these reasons, more and more peoples are health conscious and concern on the types of food they consumed. They are apprehensive not only on the appearance of the products but also the nutritional information for packaged of food selling at retails outlets.

The nutritional information or labelling has long been a mandatory in food manufacturers and retailers (King, 2006; Ophardt, 2003). Most legislation is mandating the disclosure of nutrients information of the food products. Nutritional information in fact crucially important to food process industry as it provides a means for consumers on the nutritional properties of the products. Almost all countries in this world emphasized that the nutrients contents in foods like, calorie, protein, carbohydrate, fat, sugar and others must be regulated by laboratories accredited (McVety, Ware & Lévesque 2001). In restaurant industry, food and product information on restaurant menus are normally related to quantity, quality, price, brand names, product identification, points of origin, merchandising terms, visual presentation and food preparation (Thomas & Mills, 2008a). Restaurants operators put a lot of effort to display food preparation or methods of cooking like deep fried, smoked, prepared from scratch, roasted, steamed on menus to gives customers general idea and provide an estimate time on how long food will be served and would be presented and according to Drydale & Galipue (2008) customer will look into it before make a purchase decision. Nevertheless, basic information disclosure has a relatively given minor impact on product perceptions and evaluations, unless nutritional information available to help customer better understands the contents of food they consumed (Consumer Affairs, 2008). Furthermore, in this present days customers are believed to be demanding beyond those practice and nutritional information particularly one of the importance ones to be included in the menu. Customers expect restaurateurs to provide information on the calorie, sugar, protein, carbohydrates and fat content on the menu (Thomas & Mills, 2006) and the provision of nutritional information on restaurant menu has been gaining popularity and the number of customers who are concerned on this issue has also increased (Thomas & Mills, 2008b).

The importance of nutritional information in the restaurant menu begins to catch attention among the academic scholars' from the customer perception to intention and actual purchase behaviour (Basset et. al, 2007). According to Hwang & Lorenzen (2008) providing nutritional information on restaurant menu would aid customers in making healthier choices and they were in fact willing to pay more for food that they perceive to be healthier. Burton and Creyer, 2004) claimed that customers had more positive attitudes towards the items with higher purchase intentions when the most favourable nutritional information presented on restaurant menus and having a slight negative influence and purchase intentions when unfavourable nutrition information were presented. Yamamoto et al (2005) discovered the provision of nutritional information on the menu resulted in the selection of meals with lower calorie and fat content from two out of the three restaurant menus on average and significant number of customers changed their meal orders. Adams & Ferrett (1997) investigated customer perception towards actual nutritional status of sandwiches and revealed that sandwiches which were below the standard nutrient requirement for energy were not so popular. Similar result was obtained by Cranage, Conklin & Lambert (2004) whereby the lower fat and lower calorie entrées were often chosen by the customers than the high fat and calorie content entrées when nutritional information was displayed for entrée items in a restaurant menu.

In examining the use of nutritional and ingredient information to college freshmen at the point of sales in campus dining facilities, Conklin, Lambert & Cranage (2005) found that females were more likely than males to use the nutrition information labels to make food choices. Females used nutrition information to identify and select lower fat, lower calorie foods, males used the information to select foods with higher levels of protein. The University of California's Center for Weight and Health calculated that, on an annual basis, nutritional information could reduce the average adult fast-food patron's yearly intake by 9,300 calories, preventing the equivalent of 2.7 pounds of weight gained per person per year and if customer see the nutrition information it could preventing around 40 million pounds of weight gain annually for the entire state of California (University of California, 2009). From the restaurateur perspectives, Kozup, Creyer & Burton (2003), demonstrated that some of restaurateurs also have very little knowledge on nutritional content however will to use the menu the laboratories and nutritionists assistance.

All in all, most of the research studies concluded that providing nutritional information on the menu may help customers makes healthier eating choices. Nevertheless, most of the available studies were undertaken in the developed countries with non -looking at Malaysian restaurant contextually setting. With this gap, the present study empirically investigates Malaysian customer perception on the importance of providing the nutritional information in full service restaurants menu and how it contributes to their purchase decisions. This study sought to examine the general perception of Malaysian customer toward nutritional information in the restaurant menu. In addition to compare and contrast the perception based on customer gender as well as to assess what are the most important nutrients to be included in the menu?

2. METHODOLOGY

A self reported experience through a questionnaire survey was opted in this study. The populations were among customers who had experience dining at full service restaurants. These restaurants were chosen as they are one of the fastest growing types of restaurants in Malaysia and to have received customers that are concerned about the types of food offer in the menu (Euromonitor, 2008). Shah Alam city was used as contextual setting for data collection. However, despite a large number of full service restaurant customers in this city area, only eights restaurants were given permission for the researchers of undertaking the survey in their premises with the rest refused as they do not entertain any solicitation other than business.

The survey questionnaire was divided into three major sections. Each section contained questions addressing the variables suit to the research objectives. Section A comprised five questions using nominal scale and focused on respondent demographic profile. Fifteen items were used in section B to assess the general perception of Malaysian restaurant customers toward providing nutritional information in the restaurant menu. Respondents were required to translate their view on a five type Likert scale ranging from 1 with "totally disagree" to 5 " totally agree ". Eight items in section C used to measure the importance nutrients to be included in the restaurant menu using scale ranging from 1 with "No Important at all " and 5 "Very Importance ".

The survey was undertaken on four weekends including Friday and customers were approached after they had lunch at the respective full service restaurants. The information on the anonymity and confidentiality was provided through the information sheet attached to the questionnaire. This information sheet provided the details about the researcher, the aim of the study and the purpose of the survey to be conducted. With absent of obvious problems, a total of 160 usable questionnaires were obtained.

3. RESULTS AND DISCUSSIONS

3.1. Characteristics of the Sample

55.0 % (n =88) of the respondents were female compared to 45.0 % (n = 72) male. 47 % (n= 76) were between 26 to 40 years old, followed by 33.8 % (n = 54) who were between 41 to 50 years of age as opposed 18.8% (n= 30) were under 25 years. The highest proportion of the respondents were Malays accounted for 54.4 % (n = 87) as opposed to 25.0 % (n = 40) Chinese, 17.5 % (n = 28) Indian and the smallest proportion (3.1 %, n= 5) were others. Sample also showed that 46.3 % (n= 74) of the respondents were government servants, 27.5 % (n= 44) were among the private employees, 16.9 % (n=27) students of higher institutions and 9.4 % (n= 15) were self employed. The frequency of visits by 416 Norrina Din et al. / Procedia - Social and Behavioral Sciences 42 (2012) 413 – 421 customer to the resta customer to the restaurants is also shown that 47.5 % (n= 76) dined twice a month, followed by 28.1 % (n= 45) around three times a month, 15% (n= 24) dined once a month and 9.4 % (n=15) dine four times in a month.

3.1.1. Customer Perceptions

The descriptive statistic looking at the mean score was applied in examining the respondents' perception towards nutritional information. The respondents' scores are presented in Table 1.

It is observed that majority of the customers perceived providing nutritional information in the restaurant menu is important (M= 3.82) as it can help them in determine the nutrition intake (M= 4.24). They were in agreement that interested in looking (M= 3.93) and intent to pay attention to it when choosing a menu item. These strong feelings perhaps best explained through their disagreement to the item "I could not care less whether the restaurant having nutritional information or not" with the mean score (M=2.42). It is also interesting to note that majority of restaurant customers in this study clearly expressed themselves of being knowledgeable about nutritional information (M= 3.68), having ability to comprehend nutritional information on the menu (M=3.59) and confident in using nutritional information (3.82). This is evident when they were in agreement to all these items. With these notions it was not surprising that they will always be looking for nutritional information of menu items in a restaurant in future (M= 3.97), would like to see additional nutritional information on the menu items (M= 3.95) and would not buy the products without nutritional information in the future (M=3.84). In addition, majority of the customers believed that restaurants should accurately tell nutritional information in the menu (M=4.03), not misleading them (M=3.95) and providing percentages on nutritional information could be sufficient (M=3.76). Despite these, customers totally agree that it is quite costly for restaurateur to include nutritional information in the menu. (M= 4.67).

Table 1. Showing the mean scores on the perception of restaurant customers towards nutritional information

Item No.	Items	n	Mean (*M*)	S.D
1.	Providing nutritional information in the menu is important for me	160	3.82	1.174
2.	I believed nutritional information help me determine the nutrition intake when I dine at restaurant.	160	4.24	.830
3.	I am interested in looking for nutritional information of menu items in a restaurant.	160	3.93	1.088
4.	I intent to pay attention to nutritional information while choosing a menu item in a restaurant.	160	3.77	1.047
5.	I could not care less whether the restaurant having nutritional information or not.	160	2.42	1.262
6.	I am quite knowledgeable about nutritional information	160	3.68	.975
7.	I feel confident about my ability to comprehend nutritional information on the menu.	160	3.59	1.190
8.	I am confident in using nutritional information	160	3.82	1.031
9.	I will always be looking for nutritional information of menu items in a restaurant in future.	160	3.97	.985
10.	I would like to see additional nutritional information about menu items in a restaurant in the future	160	3.95	1.089
11.	I would not buy the product without high nutritional information in the future.	160	3.84	1.186
12.	Restaurant should accurately tell nutritional information in the menu	160	4.03	.928
13.	I believe nutritional information should not be misleading.	160	3.95	1.014
14.	Percentages on nutritional information could be sufficient to provide how much of a given ingredients a food product contain.	160	3.76	1.000
15.	I believe it is quite costly for restaurateur to include nutritional information in the menu.	160	4.67	1.032

These findings indicate that customers generally having high expectations on providing nutritional information in the full service restaurant menu. This contention is in line with other researchers' findings (Thomas and Mills, 2008) that nutritional information on restaurant menu has been gaining popularity in world today as customers are getting knowledgeable, practice a better lifestyle and more health conscious particularly on eating out food.

3.1.2. Comparison Perception Based on Gender

It was envisaged that when gender was used in the instrument, there would be a greater likelihood of the female to be more concerned on nutritional information than the males' counterpart. Given this, the Independent t- test procedure was utilized to identify if there were any statistically significant differences between those responses. Table 2 reports the results from this analysis.

Out of fifteen items, ten showed statistically significant differences between females and males customers. Female customers have given greater level of agreement (M= 4.19, p = .000) than male (M=3.44) on the important of providing nutritional information in the full service restaurant menu and believe it will help them in determine the nutrition intake (M= 4.52, p= .000 compared to M= 3.95 for male). Similarly, female customers agree that they were interested in looking for nutritional information of the menu items (M= 4.19, p= .002 as opposed to M= 3.65 for male) and intent to pay attention on it while choosing a menu item in a restaurant (M= 4.05, p= 000 compared to M= 3.46 for male). With such feelings, female customers were also reported themselves as somewhat more agreeing that they were quite knowledgeable about nutritional information (M= 3.61, p= .037) than the male customers (M= 3.03), confident about their ability to comprehend nutritional information on the menu (M= 3.53, p= .009 compared to 3.04 for male) and in fact they were more confident in using nutritional information (M= 3.56, p=.011 as opposed to M=3.18 for male). Female customers' strong feelings were further supported by their somewhat higher agreement that they will always be looking for nutritional information of menu items in the future (M= 4.14, p= .000 compared to p= 3.58 for male), would like to see additional nutritional information (M=4.19, p= .000 compared to p= 3.27) and would not buy the product without nutritional information in the future (M= 4.07, p=.011 compared to M=3.09 for male). Together these points might suggest that female is slightly more concerned than the males' counterpart on certain elements related to nutritional information in the full service restaurants.

Despite slightly difference on certain aspects, both gender customers however sharing similar levels of agreement on the other items used in the instrument. This can be seen from the magnitude of the mean scores which also showing no statistically significant differences. Both male and female customers in agreement that full service restaurant should accurately tell nutritional information in the menu (M= 4.09 for female and M=4.06 for male, p= .056), should not be misleading the nutritional information (M= 4.06 for female and M= 4.04 for male, p= .805) and could be sufficient of

Table 2. Showing mean scores for items reported by male and female respondents with standard deviation, t-test results and levels of significance

No.	Items	Gender	Mean (M)	S.D	t-value	Sig 2-tail
1.	Providing nutritional information in the menu is important for me	M	3.44	1.325	-4.234	.000
		F	4.19	.868	-4.202	
2.	I believed nutritional information help me determine the nutrition intake when I dine at restaurant.	M	3.95	.938	-4.593	.000
		F	4.52	.594	-4.556	
3.	I am interested in looking for nutritional information of menu items in a restaurant.	M	3.65	1.247	-3.165	.002
		F	4.19	.838	-3.142	
4.	I intent to pay attention to nutritional information while choosing a menu item in a restaurant.	M	3.46	1.203	-3.680	.000
		F	4.05	773	-3.651	
5	I could not care less whether the restaurant having nutritional information or not.	M	2.60	.902	2.453	.055
		F	2.56	.863	2.451	
6.	I am quite knowledgeable about nutritional information	M	3.03	1.019	-1.842	.037
		F	3.61	.917	-1.838	
7.	I feel confident about my ability to comprehend nutritional information on the menu.	M	3.04	1.200	-2.652	.009
		F	3.53	1.141	-2.650	
8.	I am confident in using nutritional information	M	3.18	1.125	-1.702	.011
		F	3.56	.923	-1.696	
9.	I will always be looking for nutritional information of menu items in a restaurant in future.	M	3.58	1.063	-3.724	.000
		F	4.14	.818	-3.705	
10.	I would like to see additional nutritional information about menu items in a restaurant in the future	M	3.27	1.028	-4.005	.000
		F	4.19	.922	-3.997	
11.	I would not buy the product without nutritional information in the future.	M	3.09	1.164	-2.580	.011
		F	4.07	1.172	-2.581	
12.	Restaurant should accurately tell nutritional information in the menu	M	4.06	1.159	-2.245	.056
		F	4.09	.594	-2.221	
13.	I believe nutritional information should not be misleading.	M	4.04	1.066	-.247	.805
		F	4.06	.968	-.247	
14.	Percentages on nutritional information could be sufficient to provide how much of a given ingredients a food product contain.	M	4.01	.818	-3.712	.154
		F	4.04	.837	-3.713	
15.	I believe it is quite costly for restaurateur to include nutritional information in the menu.	M	3.76	.956	-1.547	.124
		F	3.78	.775	-1.541	

providing percentages of food product contain (M=4.04 for female and M= 4.01, p= .154 for male). To express their concerned both genders rejected the idea that they do not care less whether the restaurant having nutritional information or not. The rated mean score might explain their level of disagreement (M= 2.56 for female and M= 2.60, p= .055 for male). Analysis also revealed that male and females strongly believed and sharing the same views that quite costly for restaurateur to include nutritional information in the menu (M= 3.78 for female and M= 3.76 for males, p = .124). In sum, these results fits quite well with the idea that both male and female despite realising a slight burden for restaurateurs to place nutritional information in the menu but still demand and attentive about it.

3.1.3. The Importance Nutrients

Table 3 demonstrates the mean score and the standard deviation for each item given by customers. As can be seen, calories (M= 4.29), protein (M=4.29), fat M= 4.22) and dietary fibre (M= 4.17) appeared to be the most four important nutrients perceived by customers to be included in the menu. Majority of customers however just ascribed slightly important to nutrients like carbohydrates (M= 3.38), cholesterol (M=3.30), salt (M=3.20) and sugar (M=3.36. This finding indicates that customers generally have high expectations on the availability of at least four common nutrients if that would be possible to be included in the menu. In other words, customers probably get use to these nutrients which were normally highlighted in packaged, canned food and beverage products.

Table 3. Showing the mean scores on the important nutrients to be included in restaurant menu

Item No.	Items	n	Mean	S.D
1.	Calories	160	4.29	.756
2.	Protein	160	4.29	.772
3.	Fat	160	4.22	1.246
4	Dietary fibre	160	4.17	.818
5.	Carbohydrates	160	3.38	1.253
6.	Cholesterol	160	3.30	1.228
7.	Salt	160	3.20	1.109
8.	Sugar	160	3.36	1.167

4. IMPLICATION AND CONCLUSION

Results of this study demonstrate that there is an awaking of nutritional awareness among the Malaysian restaurant customers. Majority of restaurant customers positively perceived the important of providing nutritional information in the full service restaurant menu. Although, presently many of the Malaysian restaurant customers are believed not having such experience in the full service restaurant, providing nutritional information on restaurant menu is more likely to be accepted by them in year to come and it availability will directly influence their future purchase decision. Result also revealed that the common highlighted nutrients in food retail products like calories, protein, fat and dietary fibre are the minimum customers desire information in the restaurant menu. As with other studies, female customers were found to be slightly concerned than males customers on certain elements related to nutritional information. This is not surprising as female by nature are slightly picky than male's customers with regard to food and they are also consider by many scholars as a gatekeeper role in providing the meal and meal solution for the families.

The above scenario highlights several important points which warrant discussion. The changing in customer attitudes through educational knowledge has led to the increased in healthier eating including the restaurant food and others foodservice operation. As nutritional information on the packaged, canned food and beverage products have been used for quite sometimes; full service and other types of restaurant operators sooner or later will be facing a tremendous pressure from the customers or probably government in providing that information in the menu. This is evident when some of the established and world recognised fast restaurants in the globe including Malaysia even only on selected nutrients have started to make nutritional information available in their menu board owing to accusation of providing unhealthy food. The full service restaurant in Malaysia therefore should start to equipment themselves with nutritional knowledge as they most likely to be one of the targeting types of restaurants after the fast food despites some might argue that restaurant is not about health but all about profit.

As a conclusion, it is hoped that the recommendations and information flow from this study will facilitate restaurant operators with valuable information on the future customer concern because managing customers' needs is becoming an important and critical area as the world economy progressively turns to a service orientation. Finally, investigation on the willingness, constraints and obstacles on providing the nutritional information from Malaysian full service restaurant operators perspectives need to be further explored.

ACKNOWLEDGEMENT

Alhamdullilah, thanks to Allah for giving me the strength and blessing to complete this research and a word of thanks particularly goes to Assoc. Prof. Dr. Mohd Salehuddin b. Mohd. Zahari and Mr. Shazali b. Md Sharif, for their guidance. This research would not have been started, nor completed without the support from the Universiti Teknologi MARA (UiTM).

REFERENCES

1. Adams, A., & Ferrett,V. (1997). Customers' perception of actual nutritional status of their lunch-time sandwiches. British Food Journal, 99 (103), 71-375.

2. Basset, M.T., Dumanosky, T., Huang, C., Silver, L.D., Young, C., Nonas, C., Matte, T.D., Chideya, S., & Frieden, T.R., (2007). Purchasing behaviour and calorie information at fast food chain in New York city, American Journal of Public Health, 98 (8), 1457-1459.

3. Burton, S, and Creyer, E.H. (2004). What consumers don't know can hurt them: consumer evaluations and disease risk perceptions of restaurant menu items. Journal of Consumer Affairs, 38 (1), 121-145.

4. Conklin, M.T., Cranage, D.A., & Lambert, C.U. (2005). Nutrition information at point of selection affects food chosen by high school students. Journal of Child Nutrition & Management, 20 (2), 97-108.

5. Consumer Affairs. (2009). Many consumers misread nutrition panels. Retrieved from http://www.consumeraffairs.com /news04 /2009/06/nutrition_panel.html.

6. Cranage, D.A., Conklin, M.T., & Lambert, C.U. (2004). Effect of nutrition information in perceptions of food quality, consumption behavior and purchase intentions. Journal of Foodservice Business Research, 7, 43-61.

7. Dennison, D., Dennison, K.F., & Frank, G. C.(1994). The DINE System: Improving food choices of the public. Journal of Nutrition Education, 26(2), 87-92.

8. Drydale, J.A., & Galipue. J.A., (2008). Profitable menu planning. (4th edition). New Jersey: Pearson.

9. Euromonitor. (2008). Consumer foodservice in Malaysia. Global Market Information Database, February.

10. Gopalan, C. (1992). Undernutrition: measurement an implication. (In) Osmani S.R, (ed) Nutrition and poverty. New York: Oxford University Press: 17- 48.

11. Hwang, J., & Lorenzen., C. L.(2008). Effective nutrition labeling of restaurant menu and pricing of healthy menu. Journal of Foodservice, 19(5), 270-276.

12. King, M. (2006). Role of biology lipids. Retrieved from http://web.indstate.edu/thcme/mwking/lipids.html.

11

CHAPTER

FOOD HANDLERS' ATTITUDE TOWARDS SAFE FOOD HANDLING IN SCHOOL CANTEENS☆

Saidatul Afzan Abdul Aziz [a], Hayati Mohd Dahan [b]

[a]*Faculty of Hotel and Tourism Management, Universiti Teknologi MARA, Shah Alam 40450, Malaysia*

of Business Management, Universiti Teknologi MARA, Shah Alam 40450, Malaysia

ABSTRACT

This study measure food handlers' attitude on food safety as well as their belief and their perceive barrier toward safe food handling. Three hundred and sixty three respondents were proportionate selected randomly from four urban areas representing four zones in Peninsular Malaysia namely Alor Setar (Kedah), Kuantan (Pahang), Shah Alam (Selangor) and Melaka by using cluster sampling. This study demonstrated that food handlers' attitude have positive and significant relationships with safe food handling. It is essential for food handling certification program, which reveal behavioural change for a better quality training.

Keywords: Attitude; behavior change; food handlers; food safety training

1. INTRODUCTION

Food service industry in Malaysia is a sector targeting on local population of about 26 million persons with a short-term population of tourist and business visitors of almost 22 million persons annually (Tourism Malaysia, 2008). Food service industry faces rigid competitions, challenges as consumers demand higher quality food, government need to guarantee safe consumption of food, business owners and/or shareholders focusing on increasing technology efficiency to increase profit (Spears, 2000; USDA, 2009) that may imitate in conflicting of interest. Foodborne diseases are extensive both in developed and more so in developing countries. It is of good concern that World Health Organization (WHO, 2007) reported in the year 2005 that 1.8 million people died from diarrhoea one of vary foodborne diseases. For this reason, foodborne diseases have captured public awareness worldwide in recent years. Centre Disease Control and Prevention (CDC, 2000) identified five risk factors of food handling that add to foodborne illnesses which include improper cooking procedure, temperature abuse during storage, lack of hygiene and sanitation by food handlers, cross contamination between raw and fresh ready to-eat foods.

Students are captive customer who is usually incompetent to purchase food from external sources during six hours they are at school. It is important for schools to realize their shared responsibility to provide nutritious and safe food to their students. School meals are significant because they are one of the most perceptible instruments for policy intervention in children's diet to hold up development of healthy eating patterns (Moy, Gan and Siti Zaleha, 2006; Yabanci and Sanlier, 2007). The Ministry of Health Malaysia (MOH, 2007) reported in 2006 that the incident rate for food poisoning was 26.04 per 100,000 people. However, WHO (2008) reported in 2007 there was raise of incidence rates for food poisoning at 52.6 per 100,000 people which is one of top five diseases in Malaysia. In line with such report, there was in fact an increase in number of episodes of foodborne outbreak reported by various states in Malaysia commonly outbreaks occurring in schools (Zain and Naing, 2002; WHO, 2008;Sharif and Al-Malki, 2010).

Besides, an epidemiology study found out that since 1997, foodborne outbreaks increased 66.5% among school age group in Malaysia (Meftahuddin, 2002; Naing, Zain, Hamzah, Mat, Abdullah and Bakar, 2007). In Malaysia, food poisoning recorded the highest incidence rate of communicable disease in food and water borne disease category (44.18 per 100,000 populations) as presented in Table 1 (MOH, 2011). Previous record showed that 48% of the total food poisoning cases in year 2007 actually took place in schools with Selangor, Perlis, Kedah, Perak, Sabah and Kelantan as the six states with the most regularly reported cases (Utusan, 2008). On a smaller scope in 2008, Pahang traced more than 71% of their overall food poisoning cases in school canteen (Utusan, 2008). Further report revealed more than 88% of food poisoning cases in Johor apparently involved the students (Utusan, 2009).

These numbers established that schoolchildren have been the foremost victim in many food poisoning cases mainly in Malaysia. Children with weakened immune systems are more at risk of getting ill from food poisoning than those who are in good health. As one of many other types of foodservice operations, school canteens are most frequently cited locations for outbreaks of foodborne disease (Seaman, 2010; Sanlier and Konaklioglu, 2012). Children are most vulnerable to foodborne diseases because their immune systems are not fully developed (McSwane, Rue and Linton, 2003). The purpose of this study is to investigate the variation in attitude that influence food handlers' behavior towards safe food handling and to examine the influence of perceived barriers among food handlers behavior toward safe food handling. In the light of pressing concern pertaining to safe food consumption amongst young children, it is vital to examine the outcome on attitude of food handlers in school canteens towards safe food handling.

Table 1. Incidence rate and mortality rate of communicable disease per 100,000 populations in Malaysia

Food and Water Borne Diseases		
Communicable Diseases	Incidence Rate	Mortality Rate
Cholera	1.56	0.02
Dysentery	0.37	0
Food Poisoning	44.18	0
Typhoid	0.74	0
Hepatitis A	0.14	0

Source: Ministry of Health Malaysia (2011)

2. LITERATURE REVIEW

2.1. Attitude on food safety

Food safety is responsibility of every person who involved in food service operation to handle food. According to Nieto-Montenegro, Brown and Labarde (2008), general food handling mistakes besides serving contaminated raw food also includes inadequate cooking, heating, or reheating of food consumption of food from unsafe sources, cooling food inappropriately and allowing too much of a time lapse. Those errors might lead to food poisoning. Many studies identify the need for training and education of food handlers in public hygiene measures on microbiological food hazards, temperature ranges of refrigerators, cross contamination and personal hygiene (Ogden, 2003; Bas et. al., 2004; Worsfold and Griffith, 2010). However, in some preceding studies show no differences between staff who attended educational course with those who did not (Almanza, Namkung, Ismail and Nelson, 2007; Afifi and Abshelaibi, 2012). This statement is supporting by several studies (Compos, Cardoha, Pinheiro, Ferreira, Azevedo and Stamford, 2009; Thobaben, 2010) and it shows that although training may increase the knowledge of food safety but it might not always turn out positive change in food handlers' attitude.

Attitude is a measure of degree to which a person has favourable or unfavourable evaluation towards behaviour (Azjen, 1980). Such as person thinks that preparing and handling food hygiene is important and necessary, they are likely intend to engage the behaviour. Vladimirov

(2011) point outs the correlation of positive behaviour, attitudes and continued education of food handlers towards the maintenance of safe food handling practices. Contradict with Bas et. al. (2004) in their study establish that attitude scores of the food handlers toward foodborne diseases prevention and control was poor as well as hygiene practices scores were even low.

Social cognitive models from the area of health psychology frequently posited as important implement in improving both forecast and intervention research in safe food handling (Rennie, 1995). One such model is Health Belief Model (HBM) by Rosenstock in year 1975; which considers barriers and benefits of engaging in safe food handlings as well, as how severe food poisoning is seen to be and the degree of susceptibility to the illness. HBM consists of five variables but the researcher is focusing on two variables related to study that perceived susceptibility and barrier of taking actions. Perceived susceptibility is belief on perceived chances that one can influence by specific condition or disease (Azjen, 1980; Hilton, 2002). For example, it would be presume that more susceptible one feels to foodborne illnesses, the more critical one would foresee the outcomes. As for perceived barriers are indicative of the challenges one expects when attempting to engage in health behaviour. For example food handlers believe that washing hands would avoid foodborne illness, but they identify time constraints and lack of facilities is barrier to comply with this behaviour.

3. METHODS

Three hundred and sixty three respondents were proportionate selected randomly from four urban areas representing four zones in Peninsular Malaysia namely Alor Setar (Kedah), Kuantan (Pahang), Shah Alam (Selangor) and Melaka by using cluster sampling, resulting in response rate of 72.6%. The respondents were among food handlers in primary school canteens. This study employed selfadministered questionnaire for data collection. Respondents could attend at their own convenient time without pressure (Sekaran, 2010). The structured questionnaire is to measure all construct involved. The development of the questionnaire involves three sections. The first section is to measure attitude towards behaviours with 16 attributes developed based on past literature on perceived susceptibility and perceived barrier, and the attributes modified to suit the context of the study. The respondents were asked to rate on a seven-point Likert scale, ranging from 1 (strongly disagree) to 7 (strongly agree). The second section is to measure safe food handling with 15 attributes. For this section, the Likert scale ranging from 1 (never) to 7 (always). Finally, the third section is asking on respondents' demographic profile. The data collection period took approximately three months for completion. Data gathered were analyzed by using Descriptive statistical (mean, standard deviation and frequencies) for all variables which measure score of each item in every section that measured level of agreement of the respondents. Bivariate correlation coefficient analysis used to determine the cor safe food handling.

4. RESULTS AND DISCUSSION

4.1. Profile of respondents

According to table 2, there were 76.4% female and 23.6% male respondents. The majority of the respondents were the age group of 45 to 54 years old (26.1%). Approximately, 77.2% of the food handlers are married and majority of them are Malaysian with 92.3%. Result also showed that 68.7% education level are among secondary level with 37.4% working experience more than 8 years as food handlers. Finally, 72.3% of the food handlers had their formal training on food handling certified by the Ministry of Health, Malaysia with 93.7% had their typhoid injection. It shows that 27.7% of food handlers did not comply on food handling certification according to the Food Hygiene Regulation 2009.

Table 2. Demographic characteristic of the respondents

Demographic characteristic	Number of respondents	Percentage (%)
Gender (n=363)		
Male	86	23.6
Female	278	76.4
Age		
16 to 24	46	12.9
25 to 34	79	21.7
35 to 44	90	24.7
45 to 54	95	26.1
55 to 64	46	12.6
Over 65	7	1.9
Marital Status		
Single	73	20.1
Married	281	77.2
Others	10	2.7
Nationality		
Malaysian	335	92.3
Others	28	7.7
Education level		
None	11	3.0
Primary school	62	20.1
Secondary school	250	68.7
College or University	41	11.3
Food Handling Certification		
Yes	262	72.3
No	101	27.7
Typhoid Injection		
Yes	340	93.7
No	23	6.3
Working Experience		
6 months to 1 year	68	19.0
2 to 3 years	51	14.0
4 to 5 years	49	13.5
6 to 7 years	59	16.2
More than 8 years	136	37.4

4.2. Attitude on food safety

Respondents were asked to select their response from a Likert scale of 1 to7, where 1 = 'strongly disagree', 2 = 'disagree', 3 = 'slightly disagree', 4 = 'natural' 5 = 'slightly agree', 6 = 'agree', 7 = 'strongly agree'. Referring to table 3, majority of the food handlers agree that adhere to school canteen management guideline will reduce food poisoning occurrence (M=5.99, S.D=1.204). However, food handlers slightly agreed with controlling temperature is one of an effective way to reduce food poisoning. Therefore, food handlers agree with school cleanliness monitoring system, which provides a good feedback for improvement in food hygiene practices. Finally, majority of the food handlers agree that they should have a certified food safety training to prevent from food poisoning occurring.

Second statement is discussing on food handlers perception barrier to carry out food safety behaviour. Majority of food handlers agree that they were facing difficulties to carry out food safety behaviour if kitchen equipment are located at inconvenient place for food production and no formal food guidelines for reference if there is any mishandling occurred. In addition to this, food handlers slightly agree that small working space is barrier for them to carry out food safety behaviour. Besides, food handlers agree that without sufficient dry and wet storage influence their food safety practices. As such, food handlers agree that lack of supervisor commitment and inadequate training pertaining to safe food handling will lead to food poisoning occurrence. This explains the need for relevant training, as reported above only 77% was given such training. Food handlers moderately agree that they face barriers to carry out food safety behavior especially with unavailability of food handling guideline, kitchen equipment in an inconvenient location and small working space to prepare food. This analysis revealed food handlers beliefs the importance of safe food handling and there are barriers to carry out food safety behaviour, but the food handlers agreed that they must have positive attitude on handling safe food (Mitchell, Fraser and Bearou, 2007).

4.3. Relationship between attitude and safe food handling

The relationship between IV and DV are weak but positively correlated. The result of the correlation coefficient between compliance and safe food handling is $r = .065, p>.01$, whilst perceived barrier and safe food handling is $r = .084, p>0.01$. From this outcome, food handlers have positive attitude towards compliance to safe food handling but are not provided with school canteen guidelines to guide safe hygiene practices. Perceived barrier are not considered a hindrance to safe food handling but again relevant guidelines are not made available to ensure good kitchen practice. Sharif and Al-Malki (2010) reported that beliefs on food safety and perceived barrier are important predictors to ensure food handlers comply with safe food handling. But this study found weak correlation perhaps because this study is conducted in cities, where food handlers are more exposed. However the biggest hindrance to safe food handling are the unavailability of relevant printed material that provide guidance to canteen food handling hygiene and good kitchen practices (Both these items scored mean score of 6.00 (agreed) for Guidance to canteen food handling and 5.00 (moderately agree) for guide to food handling hygiene.

Table 3. Mean attitude on food safety of sampled food handlers

Statement	Mean (M)	Standard Deviation (S.D)
Beliefs		
Adherence to School Canteen Management Guidelines will reduce chances of food poisoning occurrence	5.99	1.204
Temperature Controls are an effective method of reducing the number of cases of food poisoning	5.32	1.398
Monitor temperature of raw and cooked food are vital	5.70	1.311
School cleanliness monitoring system provides good feedback for improvement	5.80	1.212
All food handlers should have a food safety training qualification	5.83	1.267
Food Act 1983 enhances compliance to food hygiene practices	5.97	1.110
Perceived Belief		
It is difficult for me to carry out food safety behaviour due to...		
Small working space to prepare food	4.45	1.927
Unavailability of food handling guideline	4.56	1.900
Kitchen equipment in an inconvenient located in	4.48	1.981
Insufficient kitchen equipment	4.26	1.981
Insufficient dry storage	4.02	2.045
Insufficient wet storage	4.17	2.026
Inadequate training pertaining to safe food handling	4.29	1.877
Lack of team work	4.14	2.002
Lack of supervisor commitment to safe food handling	4.23	1.921

Table 4. Correlation between attitude and safe food handling

		Food Handling	Compliance	Perceived Barrier
Food Handling	Pearson Correlation	1	.065	.084
	Sig. (2-tailed)		.215	.108
	N	363		
Compliance	Pearson Correlation	.065	1	-.064
	Sig. (2-tailed)	.215		.223
	N	363		
Perceived Barrier	Pearson Correlation	.084	-.064	1
	Sig. (2-tailed)	.108	.223	
	N	363		

5. CONCLUSION

Attitude toward compliance is positively correlated to safe food handling especially in terms of adherence to school canteen management guideline and Food Act 1983 regulation in enhancing compliance with food handling. Findings in this study reported that adherence to school canteen rules and government regulations play important roles in influencing food handlers' attitude towards safe food handling. It was so found that Perceived barrier generally do not hinder safe food handling. However, small working space, inconvenient location for kitchen equipment and unavailability of food handling guideline toward achieving safe food handling are individual items in the questionnaire that scored high, showing respondents' agreement to these as barriers in safe food handling. In conclusion, proper printed guidelines on canteen food handling and a separate guideline on kitchen good practice may have an impact on safe food handling behavior. This finding is hoped contribute to Malaysia's effort to avoid or reduce food poisoning occurrence amongst schoolchildren in the future.

REFERENCES

1. Afifi, H. S. & Abushelaibi, A. A. (2012). Assessment of personal hygiene knowledge, and practices in Al Ain, United Arab Emirates. *Food Control*, 25(1), 249 253.

2. Almanza, B. A, Namkung, Y., Ismail, J. A. & Nelson, D. C. (2007). Clients' safe food-handling knowledge and risk behavior in a home-delivered meal program. *Journal Of The American Dietetic Association*, 107(5), 816 -21.

3. Ansari-Lari, M., Soodbakhsh, S. & Lakzadeh, L. (2010). Knowledge, attitudes and practices of workers on food hygienic practices in meat processing plants in fars, Iran. *Food Control*, 21, 260-263.

4. Azjen, I. & Fishbein, M. (1980). *Understanding attitudes and predicting social behavior*. Prentice Hall Inc.

5. Bas, M., Ersun, A. S. & Kivanc, G. (2006). The evaluation of fod hygiene knowledge, attitudes and practices of food handlers' in food businesses in Turkey. *Food Control,* 17, 317-322.

6. Belot, M. & James, J. (2009). *Healthy school meals and educational outcomes.* Economic & social research council. ISER working paper series. www.iser.essex.ac.uk.

7. Bourke, J. R. & Bates, E. J. (2002). Health concerns 'drive' foods service packaging. New Foofd, 5(3), 80-2.

8. Centers for Disease Control and Prevention. (2000). *Surveillance for foodborne disease outbreak United States, 1993-1997.*

9. Chen, M.F. & Li, H. L. (2007). The consumer's attitude toward genetically modified foods in Taiwan. *Food Qualiy and Preferences,*18, 662-74.

10. Clayton, D. A. & Griffith, C. J. (2004). Observation of food safety practices in catering using notational analysis. *British Food Journal,* 106(3), 211-227.

11. Calyton, D. A., Griffith, C. J., Price, P. & Peters, A. C. (2002). Food Handlers' beliefs and self-reported practices. *International Journal of Environmental Health Research,* 12, 25-39.

12. Codex Alimentarius Commission, Joint FAO/WHO Standards Programme. (2003). *Food hygiene: Basic texts.* 2nd Edition.

13. Compos, A. K. C., Cardonha, A. M. S., Pinheiro, L. B. G., Ferreira, N. R., Azevedo, P. R. M. & Stamford, T. L. M. (2009). Assessment of personal hygiene and practices of food handlers in municipal public schools of Natal, Brazil. *Food Control,* 20, 807-810.

14. Ehiri, J. E., Morris, G. P. & McEwen, J. (1997). Evaluation of a food hygiene training Course in Scotland. *Food Control,* 8, 137- 147.

15. Food and Agriculture Organization/World Health Organization. (2004). *Regional conference on food safety for the Asia and the Pacific, Seremban.* Retrieved from www.fao.org/DOCREP/ MEETING/006/AD698E.HTM.

16. Food Quality Control Division by Ministry of Health Malaysia. (2009) Food *Act 1983 (Food Hygiene Regulations, 2009)*. Retrieved from http://fsq.moh.gov.my/modules.

17. Herztman, J. & Barrash, D. (2007). An assessment of food safety knowledge and practices of catering employees. *British Food Journal,* 109(7), 562-576.

18. Hilton, J. (2002). Reducing foodborne disease: Meeting the food standards agency's targets. *Nutrition & Food Science,* 32(2), 46-50.

19. Ko, W. H. (2010). Evaluating food safety perceptions and practices for agricultural food handler. *Food Control,* 21, 450-455.

20. McSwane, D., Rue, N., & Linton, R. (2003). *Essentials of food safety and sanitation*. Third Edition. Pearson Education. New Jersey.

21. Meftahuddin, T. (2002). Review of the trends and causes of food borne outbreaks in Malaysia from 1988 to 1997. *Medical Journal Malaysia,* 57(1), 70-9.

22. Ministry of Education, Malaysia. (2008). *Number of government assisted school 2003-2007*. Retrieved from: http://www.moe.gov.

23. Ministry of Health, Malaysia. (2007). *Incidence rate and mortality rate of communicable disease per 100,000 population in 2006*. Retrieved from http://www.moh.gov.

24. Mitchell, R. E., Fraser, A. M. & Bearon, L. B. (2007). Preventing food-borne illness in food service establishments: Broadening the framework for intervention and research on safe food handling behaviors. *International of Environmental Health Research,* 17(1), 9-24.

25. Moy, F.M., Gan, C. Y. & Siti Zaleha M. K. (2006). Eating patterns of school children and Adolescents in Kuala Lumpur. *Malaysia Journal Nutrition,* 12 (1),1-10.

26. Naing, N. N., Zain, M. M., Hamzah, W. M., Mat, H.A., Abdullah, N. & Bakar, A. M. H. (2007). A study on effective of health educatin program on knowledge, attitude and practice (KAP) of food

handlers towards food-borne disease and food safety. *International medical journal Tokyo,* 14(4), 253-260.

27. Namkung, Y. & Jang, S., (2007). Does food quality really matter in restaurants? Its Impact on customer satisfaction and behavioral intentions. *Journal of Hospitality Management.* 27(1), 23-29.

28. Nieto-Montenegro, S., Brown, J. L. & Laborde, L. F. (2008). Development and assessment of pilot food safety educational materials and training strategies for hispanic workers in the mushroom industry using the health action model. *Food Control,* 19(6), 616 633.

29. Niode, O., Bruhn, C. & Simonne, A. H. (2011). Insight into Asian and hispanic restaurant manager needs for safe food handling. *Food Control,* 22(1), 34 42.

30. Ogden, J. (2003). Some problems with social cognition models: A pragmatic and conceptual analysis. *Health Psychology,* 22(4), 424 428.

31. Olsen, S. J., MacKinon, L. C., Goulding, J. S., Bean, N. H. & Slutsker, L. (2000). *Surveillance for food-borne disease outbreak in United States, 1993-1997.* Retrieved from: http//www. Cdc.gov/epc/ mmwr/preview/mmwrhtml/ss4901a1.htm.

32. Osaili, T. M., Obeidat, B. A., Abu Jamous, D. O. & Bawadi, H. A. (2011). Food safety knowledge and practices among college female students in North Of Jordan. *Food Control,* 22(2), 269 276.

33. Paulhus, D. L. (1998). Interpersonal and intrapsychic adaptiveness of trait self enhancement: A mixed blessing? *Journal Of Personality And Social Psychology,* 74, 1197-1208.

34. Payne-Palacio, J. & Theis, M. (2001). *Introduction to foodservice.* 9th edition. Prentice Hall Inc.

35. Redmond, E. C. & Griffith, C. J. (2003). Consumer food handling in the home: a review of consumer food safety studies. *Journal of Food Protection,* 1(1), 130-61.

36. Reid, A., Wood, D. & Kinney, D. (1998). Food hygiene information: Power to the people? *Nutrition & Food Science*, 98(3), 138 144.

37. Rennie, D. M. (1995). Health education models and food hygiene education. *The Journal Of The Royal Society For The Promotion Of Health*, 115(2), 75 79.

38. Rodgers, S. (2005). Applied research and educational needs in food service management. *International Journal of Contemporary Hospitality Management,* 17 (4)304-314.

39. Ruth, M. W. Y. & Morris, J. (2001). food safety risk. *British Food Journal,*103,170-187.

40. Sanlier, N. & Konaklioglu, E. (2012). Food safety knowledge, attitude and food handling practices of students. *British Food Journal, 114*(4), 469 480.

41. Sahimi, N.N. & Said, I. (2012). Children selections of the physical elements in the preschool environment. *Journal of ASIAN Behavioural Studies*, 1(2), 69-79.

42. Seaman, P. (2010). Food hygiene training: Introducing the food hygiene training model. *Food Control*, 21(4), 381 387.

43. Seaman, P. & Eves, A. (2006). The management of food safety- the role of food hygiene training in the UK service sector. *Hospitality Management,* 25, 278-296.

44. Selamat, J., Shamsudin, M. N. & Dullati, M. S. (2002). *Food Outlook: Malaysia. Pacific Food System Outlook 2002-2003*, Pacific Economic Cooperation Councils. http://www.pecc.org/food

45. Sharif, L. & Al-Malki, T. (2010). Knowledge, attitude and practice of taif university students on food poisoning. *Food Control,* 21, 55-60.

46. Singh, P., Fook, C.Y. & Sidhu, G. K., (2008). *A comprehensive guide to writing a research proposal*. Venton Publishing.

47. Spears, M.C. (2000) *Foodservice Organizations: A managerial and systems approach*. Fourth Edition. Prentice Hall. New Jersey.

48. Thobaben, M. (2010). Causes and prevention of foodborne illness. *Home Health Care Management & Practice*, 22(7), 533 535.

49. Thomson, J. S., Radhakrishna, R. & Bagdonis, J. (2011). Extension educators' perceptions of the local food system. *Journal Of Extension*, 49(4), 1 13.

50. Tokuc, B., Ekuklu, G., Berberoglu, U., Bilge, E. & Dedeler, H. (2009). Knowledge, attitudes and self-reported of food service staff regarding food hygiene in Edirne, Turkey. *Food Control,* 20, 565-568.

51. Tourism Malaysia. (2008). *Malaysia key performance indicator 2008*. Retrieved from http://www.tourism.gov.my/ corporate/ research. asp?page=facts_figures.e

52. U.S Food and Drug Administration. (2004). *FDA Report on the occurrence of foodborne illness risk factors in selected institutional foodservice, restaurant, and retail food store facility types*.Retrieved from: http://www.cfsan.fda.gov/~dms/retrs2c.html.

53. United States Department of Agriculture. (2009). *Malaysia HRI food service sector food service Malaysia Annual Report. Global Agriculture Information Network (GAIN) Report*. Retrieved from www.fas.usda.gov/gainfile/200901/146326991.pdf.

54. Veiros, M. B., Proenca, R. P. C., Santos, M. C. T., Smith, L. K., & Rocha, A. (2009). Food safety practices in a portuguese canteen. *Food Control,* 20, 936-941.

55. Vemula, S. R., Kumar, R. N., & Polasa, K. (2012). Foodborne diseases in India A review. *British Food Journal*, 114(5).

56. Vladimirov, Z. (2011). Implementation of food safety management system in bulgaria. *British Food Journal*, 113(1), 50 65.

57. World Health Organization, (2008). *Malaysia health situation and trend*. Retrieved from www.wpro.who.int/ countries/2008/maa/ health_situation.htm.

58. World Health Organization. (2007). *Food safety and foodborne illness.* Retrieved from: http://www.who.int/mediac entre/ factsheets/fs237/ en/print.html

59. Worsfold, D., & Griffith, C. (2010). Experiences and perceptions of secondary food hygiene training: A preliminary study of five larger catering companies in South East Wales. *Perspectives In Public Health*, 130(4), 173 179.

60. Worsfold, D., & Griffith, C. J. (2003). A survey of food hygiene and safety training in the retail and catering industry. *Nutrition & Food Science*, 33(2), 68 79.

61. Yabanci, N., & Sanlier, N. (2007). An assessment of hygienic conditions in nursery schools' kitchens. *Nutrition & Food Science,* 37,(6),419-426.

62. York, V. K., Brannon, L. A, Shanklin, C. W., Roberts, K. R., Howells, A. D., & Barrett, E. B. (2009). Foodservice employees benefit from interventions targeting barriers to food safety. *Journal Of The American Dietetic Association*, 109(9), 1576 81.

63. Zain, M. M., & Naing, N. N., (2002). Southeast Asian med public health socio-demographic characteristics of food handlers and their knowledge , attitude and practice towards food sanitation, 33(2).

64. Utusan Online, (2011, March 30). 50% kes keracunan makanan berlaku di kantin sekolah. Retrieved April 8, 2013, fromwww.utusan.com.my

65. Utusan Online, (2009), Nov 16). Majoriti kes keracunan babitkan pelajar. Retrieved May 10, 2013, from www.utusan.com.my.

66. Ministry of Health (2011). Health facts 2010. Retrieved June 21, 2013 from www.moh.com.my.

67. Pallant, J. (2005). *SPSS survival manual: A step by step guide to data analysis using SPSS for windows version 12 (2nd Ed.).*Sydney; Allen and Unwin.

12

CHAPTER

FOOD SAFETY TRACEABILITY SYSTEMS IN THE MARITIME CATERING LOGISTICS

Basem Azmy Saad Boutros[1], Mahmoud Mahmoud Hewedi[1], Kevin R. Roberts[2], Farida M. Megahid[3]

[1]*Hotel Studies Department, Faculty of Tourism and Hotels, Fayoum University, Fayoum, Egypt*

[2]*Department of Hospitality Management and Dietetics, College of Human Ecology, Kansas State University, Manhattan, USA*

[3]*Hotel Studies Department, Faculty of Tourism and Hotels, Menoufia University, Sadat City, Egypt*

ABSTRACT

This paper aims to explore food safety traceability in the onboard food and beverage operations on cruise ships through a case study approach. Semi-structured interviews were conducted with galley and foodservice staff. Additionally, a self-administered questionnaire was designed and distributed to logistics managers and employees of three Egyptian cruise companies. Results indicated that cruise ships in this study use paper-based and manually-entered (electronically stored) data traceability systems. Managing food safety by effective product recall, complying with legislation, and improving product quality are among the most important perceived benefits identified by foodservice staff. Maintaining a good reputation, ensuring food safety, reducing foodborne illnesses, suppliers' commitment to quality specifications, and maintaining guest confidence are considered the most important benefits from implementing a food traceability system. The results also depicted that the challenges managers face implementing or improving a food traceability system included high cost, absence of unified traceability standards between cruise companies and food suppliers, shortage of skilled staff, and lack of knowledge and satisfaction with current foodservice operation systems onboard cruise ships. Other findings elucidated that the critical success factors for implementing traceability included strict food traceability legislation, ensuring the adequacy and correctness of traceability information, adopting a food traceability system by the cruise company management, the standardization of food traceability information identification, and technology support for food traceability systems implementation from government. Other factors were reported of less importance. This research concluded the urgent need for more cooperation between cruise companies, food suppliers, and relevant governmental authorities to adopt food traceability systems and the need of authorizing and setting compulsory standards and regulations.

Keywords: Food Safety Traceability, Cruise Ships, Traceability Systems, Logistics

1. INTRODUCTION

1.1. Cruise Tourism and Related Foodborne Diseases

In recent years, traveling by sea excursion has become popular. Cruises were first offered through steamship companies from main United States ports, such as New York and Canaveral, as an integrated service which combined sea passage and hotel services [1]. Vogal, Papathanassis, and Wolber [2] indicated that cruising was a fast growing segment of the tourism industry and the innovation in services provided onboard cruise ships had made cruises an attractive vacation option. The package vacations that cruises offer make them one of the most popular tourism products [3][4] . Véronneau and Roy [5] reported that the cruise industry carried 11.5 million passengers around the world in 2005, and it is estimated to reach 21.7 million by the end of 2014 [6] . Furthermore, North America and Asia have the biggest share of cruise tourism markets [7] .

The growth in the cruise industry and the nature of cruise ships as a semi-closed environment in which people live, eat, and drink from common food and water supplies may facilitate the spread of infectious diseases [8] . Mouchtouri et al. [9] , stated that features on the ships' water systems, including bunkering from ports, desalination, storage, and proximity to non-potable water systems, could increase the risk rate of contamination. Therefore, cruise ships may be potential sites for disease outbreaks and in particular foodborne and waterborne diseases. Waterman [10] discussed the first well-documented outbreak of Staphylococcal food poisoning onboard a cruise ship. Several studies have reported gastrointestinal outbreaks onboard cruise ships, including E-coli, Norovirus, and Legionnaires' disease, all resulted from food and/or water served onboard cruise ships [11] -[15] . The World Health Organization [7] has noted that many foodborne outbreak incidents on the affected cruise ships are ascribed to cross contamination, use of contaminated raw food, and infected food handlers and indicated that many outbreaks are never informed or told to health authorities and only a small percentage of them are published. Therefore, The United Nations World Tourism Organization [16] indicated that the cruise ship sector required more in-depth studies due to its growth, economic impact, and complexity.

In response to several outbreaks onboard ships, the Vessel Sanitation Program was established by the United States Public Health Service's Center for Disease Control and Prevention in 1975. The goal of the Vessel Sanitation Program was to minimize the potential for gastrointestinal illnesses and it has helped to reduce the rate of diseases outbreaks on ships [5] [8] .

1.2. The Emergence of Food Traceability Concept and Potential Benefits

Consumers demand a food supply where food safety can be documented and tracked during every stage of production, processing, distribution, and service. Folinas [17] , Regattieri [18] , and Wu et al. [19] claimed that due to the emergence of Bovine Spongiform Encephalopathy, traceability has become an essential tool for food safety, quality, and tracking within the meat supply chain. Moreover, product quality improvement, need for healthy consumption, reducing risk rate and liability, and brand name protection are among the most common driving forces for traceability system implementation [20] . From a legislative point of view, Engelseth [21] stated that the European Union General Food Law of 2005 required that food and feed business operators identify the immediate supplier of the product in question and the immediate subsequent recipient, which has increased the demand for tractability systems in the foodservice industry.

The foodservice industry has addressed the management of food hygiene, safety, and quality through the introduction of Hazard Analysis and Critical Control Points (HACCP), ISO 9001, and other initiatives. However, Sperber [22] argued that HACCP systemsare not sufficient to ensure food safety in the entire food supply chain when applied alone, it must be supported by prerequisite programs and systems, such as product traceability, feed contamination prevention, and assuring containers and packaging materials safety. Matsuda [23] claimed that food safety procedures alone cannot improve consumer confidence without conveying appropriate information of the product through the implementation of traceability. Many authors have defined the term traceability, but most have focused on the same core concept. For instance, The Food Marketing Research and Information Center [24] defines traceability as "the ability to follow the movement of food products or ingredients through specified stages of production, processing, and distribution".

With regard to the application of traceability systems in foodservice, Smith et al. [25] postulated that traceability system would be introduced in the foodservice industry in the near future. Golan et al. [26] noted that some restaurants and grocery stores now require their suppliers to adopt traceability systems, which must be verified by a third-party verification process.

Given that the flowing nature of onboard logistics operation on cruise ships, Véronneau and Roy [5] maintained that sanitation and sourcing of product should be of high importance. Tracking produce and supplies and holding suppliers accountable is critical in ensuring the safety of products onboard. Several recent studies [20] [25] [27] -[30] have contended that there are direct benefits from implementing food traceability systems, including supply chain optimization, enhancing food safety and quality, ensuring product authenticity, complying with food safety legislation, differentiation and competitive business ad-vantages, and responding to specific needs such as consumers with food allergies.

1.3. Food Traceability Framework and Technical Instruments

While many models have been introduced to support traceability information systems, each model views the issue from a different perspective. Food traceability systems are based on four elements, product identification and process linking, data to trace, product routing and data retrieval, and traceability tools [3] [18] [31] . The UK Food Standard Agency [27] indicated that products and processes were the key components of a traceability system. Depending on the complexity of the supply chain, the traceability system can be either paper-based or Information Technology (IT) based, which depends on a traceability information carrier that stores and carries related information of a food product. In this concern, optical systems like barcodes, Radio Frequency Identification (RFID), and laser etching on edible labels for fruits and vegetables have been implemented in the food industry [29] [32] -[35] .

The majority of literature relating to food traceability focuses on using instruments and potential advantages of adopting a traceability system. There is a paucity of research related to the traceability of food served within a cruise ships' foodservice operations. Therefore, the purpose of this study is to explore food safety traceability systems in the onboard food and beverage operations of passenger' ships. Specific objectives are to:

1. Identify the current traceability status with regard to coordination with food supply companies.

2. Investigate the potential benefits from implementing food safety traceability systems in the onboard logistics.

3. Determine the challenges face implementing food safety traceability systems schemes.
4. Identify the existing success factors for implementing food safety traceability systems.

2. METHODOLOGY

2.1. Design of the Case Study

The case study method was used for this research paper. This method depends on a detailed investigation through data collection from a limited sample to understand the complex relationships that exist. The reason for choosing the case study technique was to obtain a detailed description about the phenomenon under investigation and to understand the features of the case in the real life. According to Finn, Walton, and Elliott-White [36] , the case study is based on a combination of data collection methods to improve the validity of the research. Therefore, qualitative and quantitative techniques were adopted in the current study. The data were collected using a self-administered questionnaire that is adopted and developed from previous studies [20] [37] [38] . The questionnaire was prepared, piloted, and distributed to all logistics managers and employees of the investigated cruise ships to produce quantitative data for analysis.

Semi-structured interviews were carried out with galley and foodservice staff onboard cruise ships to identify their views on the potential benefits of implementing an internal food traceability system onboard the ship and the challenges that may affect their implementation of internal food traceability. The interview framework questions are summarized in Figure 1. All interviews were transcribed. An interview lasted, on average, 30 minutes. The study protocol was approved by the universities' institutional review board.

Figure 1. A framework of interview questions.

2.2. Population of the Study and Response Rate

The study domain includes three Egyptian cruise line companies that were selected as a convenience sample. The questionnaire was distributed from February to September 2013 to all logistics mangers and staff, only five questionnaires out of 20 distributed was returned. Although the response rate was lower than anticipated, because this study was concerned with developing a thorough understanding of each company's practices and not generalizing the sample to the population, five valid questionnaires were considered sufficient. In addition, three interviews were held with foodservice and galley staff.

3. RESULTS AND DISCUSSION

3.1. The Concept of Food Traceability

All respondents indicated that they know the meaning of the term traceability in the context of the ship's food supply. However, their definitions were limited to the purchasing and storing processes, without dealing with food traceability as it flows through the ship. According to the definitions outlined by the Food Marketing Research and Information Center [24], none of their definitions expressed the proper meaning and purpose of traceability. That may be due to the fact that the concept of traceability is considered new and the attention to food traceability began only few years ago as it was reported by Senneset, Forås, and Fremme [31] .

3.2. Current Traceability Status

Results from the respondents at the three cruise line companies show that they are currently using a paper-based traceability and manually-entered (electronically stored) data systems. The reason why these companies implement such traceability systems may be because of other advanced electronic systems are costly and not suitable for small businesses as previously found [39] . Other reasons could be lack of knowledge about advanced food traceability systems.

All respondents in the three companies described their traceability system as somewhat satisfactory except one respondent who described it as very satisfactory. Interestingly, all respondents expressed their willingness to change for more advanced traceability systems as a result of explaining to them how this could be beneficiary to their business. Respondents at Company I and Company II want to use bar codes while the respondents at company III would like to change to use a Radio Frequency Identification system (RFID).

3.3. Potential Benefits of Implementing a Food Traceability System

The mean scores for the potential benefits of implementing food traceability system are shown inTable1 The three cruise companies investigated deem protecting food safety by effective product recall procedures, complying with legislation, and improving product quality as the most important benefits of traceability. Concerning compliance with legislation, the results agree with what previously stated by Xiong et al. [30] as traceability systems or process could meet each country's food safety legislation and serve the public health. However, the food safety legislation in Egypt does not require a mandatory traceability system and the traceability measures are only addressed to exported food produce through an ongoing project of The Egyptian Traceability Centre for Agro-Industrial Exports. Reducing the risk rate of foodborne disease outbreaks ranked the second potential benefit. Other perceived benefits include ensuring food authenticity, differentiation, providing nutrition and allergy information ranked the third. Reducing production and labor costs ranked the fourth and the fifth respectively as of minor importance. These results agree to some extent with previously reported results [20] [29] [40] .

Table 1. Potential benefits of food traceability systems.

Potential benefits	**Mean**	*Rank*
Protecting food safety by effective product recall.	4.8	1
Complying with relevant legislation.	4.8	1
Improving product quality.	4.8	1
Reducing risks associated with food-borne diseases outbreaks.	4.6	2
Differentiating your foodservice operation.	4.4	3
Ensuring food authenticity.	4.4	3
Providing nutrition and allergy information for passengers.	4.4	3

3.4. The Challenges of Implementing a Food Traceability System

The results clearly indicated the challenges of implementing a food traceability system include the high initial cost, the absence of unified traceability standards between cruise companies and food suppliers, and the lack of skilled staff and are considered the most important difficulties for adopting a traceability system. Other challenges of minor importance include limited staff time for additional tasks other than their regular formal duties, lack of governmental support, less flexibility in the foodservice operation, no enforcement for recording traceability information, and uncertainty about future benefits. These results are similar to what previously found by Zhang et al. [38] .

Chrysochou, Chryssochoidis, and Kehagia [29] indicated that the implementation of advanced traceability systems like Radio Frequency Identification (RFID) is still limited due to the high initial investment. On the one hand, the lack of standardization and the diversity of the traceability systems, resulting in incompatibility between foodservice operations and suppliers in the food supply chain [18] [41] is still a critical issue. On the other hand, the Food Standard Agency [27] has noted that foodservice operations will be apprehensive about investing in a traceability system until they are convinced about its viability. Moreover, Engelseth [21] demonstrated that regulations by governments in terms of demanding food traceability would compel businesses to track food products.

3.5. Factors of Success for Implementing a Food Traceability System

The mean scores for the perceived success factors for implementing a food traceability system are shown in Table2 The mean for nine of the 10 variables were between 3.4 - 4.8; hence, these factors are considered essential for implementing a traceability system. The factor linked to passengers' willingness to pay an increased price for food that is traceable, had a mean value of 2.60. The two highest crucial factors for the success of implementing food traceability system are "Strict food traceability legislation" and "Ensuring the adequacy and correctness of traceability information". Our findings agree with what was previously reported by Miao [37] as the authenticity of traceability information affects the quality of information and the trust between cruise companies and food suppliers. Moreover, strict laws with regard to traceability are critical, especially in Egypt where traceability is at its early stage and only implemented in food exportation. The factors entitled "senior management requiring a food traceability system", "the standardization of traceability information", and "technology support from government" are in the second rank and have the mean of 4.6.

Table 2. Success factors for implementing a food traceability system.

	Factors	Mean	Rank
1.	Strict food traceability legislation.	4.8	1
2.	Ensuring the adequacy and correctness of traceability information.	4.8	1
3.	Senior management requiring a food traceability system.	4.6	2
4.	The standardization of food traceability information identification.	4.6	2
5.	Technology support for food traceability systems implementation from government.	4.6	2
6.	Increased education and training for employees about traceability systems.	4.4	3
7.	Full understanding of food traceability from passengers.	3.8	4
8.	Government funding for food traceability implementation.	3.6	5
9.	Sharing information between the passenger ships and food supply companies.	3.4	6
10.	Passengers' willingness to pay an increased price for traceable food.	2.6	7

The mean of the factor named "increased education and training for employees about traceability systems" is 4.4 and is 3.8 for that is called "full understanding of food traceability by passengers". This finding agrees with Engelseth [21] who asserted that employees' participation in developing a food traceability system is considered the cornerstone for establishing traceability. Van Rijswijk et al.[42] as well, stated that consumers' perception of food traceability is influenced by conveying other quality assurance information through certified traceable food.

Matsuda [23] stressed on that the role of government in supporting traceability has not just to make laws, but rather providing financial assistance in establishing, developing and promoting traceability procedures. The mean of factor "passengers' willingness to pay an increased price for traceable food" valued only 2, 6 how far is it comparable with similar results. In this respect, Mai et al. [20] reported that bringing traceability and quality assurance together increases the consumer willingness to pay for traceable food.

3.6. Interview Analysis

The galley and foodservice staff interviews consisted of seven questions about how they would define food traceability, what internal traceability procedures they were familiar with, what potential benefits they perceived of a traceability system, challenges they face in the implementation of traceability systems and their perceived critical success factors.

Results showed that none of the respondents is aware of the food traceability concept. Ensuring food safety through purchasing and storing processes is the most dominant definition of food traceability represented by the respondents. However, one respondent defined traceability as ensuring appearance quality characteristics of food. Companies subject to the current research use paper based internal system to track food from storerooms to the kitchen. Concerning the potential benefits of implementing traceability; maintaining a good reputation, ensuring food safety, reducing food poisoning and suppliers' commitment to quality specifications were found among the most potential benefits seen by respondents. These results are in good agreement with the previous findings of the questionnaire analysis and Mai et al. [20] and Chrysochou, Chryssochoidis, and Kehagia [29] . Ensuring food safety and maintaining guest confidence were also found as potential benefits for passengers. On the other hand, lack of knowledge, dissatisfaction with the foodservice operation system and high cost onboard of the investigated ships, are respectively the most apparent challenges viewed by interviewee to the implementation of a food traceability system. Apparently, educating and training of employees, management commitment to adopt a food traceability system, and securing financial needs, equipment and technology were found among the most important factors for the success of demanding a traceability system according to the respondents.

4. CONCLUSIONS

This paper aims to explore food traceability systems on passenger cruise ships through a case study approach. The focus of a case study is to understand the topic of interest, not to generalize to other populations. Topics explored included identifying the current traceability systems, investigating perceived benefits of implementing food traceability systems, determining if there are challenges to implementing food traceability systems, and identifying the most crucial success factors for implementing food traceability systems in the onboard logistics in cruise ships.

Paper-based traceability systems and manually-entered (electronically stored) data traceability systems were found as the only systems used on the investigated cruise ships. Effective product recall, complying with legislation, and improving product quality were the most important benefits of implementing the traceability. The findings are supported by the previously reported results by previous researchers [20] [29] [30] [40] . The obtained results from the interview showed that maintaining a good reputation, ensuring food safety, reducing food poisoning and suppliers' commitment to quality specifications, and maintaining guest confidence are considered the most important benefits from implementing an internal food traceability system on the cruise ship.

The study also found that the high cost of applications, absence of unified traceability standards between cruise companies and food suppliers, and lack of skilled staff are considered the important difficulties for adopting traceability systems. Interview results indicate that lack of knowledge and satisfaction with current foodservice operation system onboard the investigated ships are the most important challenges for implementing a food traceability system. These findings are supported by research conducted by Zhang et al. [38] .

Strict food traceability legislation, ensuring the adequacy and correctness of traceability information, senior management requiring a food traceability system, the standardization of traceability information, and technology support from government were found respectively according to their importance as the most critical success factors for implementing a traceability system onboard cruise ships. According to interview results, educating and training of employees, management's commitment to adopt a food traceability system, and securing financial needs, equipment, and technology were found as the most important factors for implementing an internal food traceability system on the investigated cruise ships.

The findings support the need for more cooperation and interlinking between cruise companies, food suppliers, and relevant governmental authorities to implement food traceability. In addition, the study recommends that more efforts could be put in place by the government to encourage cruise ships to adopt food traceability systems and making them aware of the different systems that could be compatible with them. Future researches are needed to focus on the cost-benefit analysis of using traceability in cruise ships, especially internally. Moreover, the extent to which cruise companies accept and implement traceability systems needs more investigation. Perception and interest of guests related to traceability could be also another subsequent for future research.

REFERENCES

1. Dickinson, B. and Vladimir, A. (2008) Selling the Sea: An Inside Look at the Cruise Industry. 2nd Edition, John Wiley & Sons, Inc., Hoboken, 68-73.
2. Vogel, M., Papathanassis, A. and Wolber, B. (2012) The Business and Management of Ocean Cruises. 1st Edition, CAB International, Massachusetts, 3.
3. Lois, P., Wang, J., Wall, A. and Ruxton, T. (2004) Formal Safety Assessment of Cruise Ships. Tourism Management, 25, 93-109.
4. Gibson, P. (2006) Cruise Operations Management. 1st Edition, Butterworth-Heinemann, Massachusetts, 20.
5. Véronneau, S. and Roy, J. (2009) Global Service Supply Chains: An Empirical Study of Current Practices and Challenges of a Cruise Line Corporation. Tourism Management, 30, 128-139.

6. Cruise Lines International Association (2014) CLIA 2014 State of the Cruise Industry Report. CLIA, New York, 1.

7. The World Health Organization (2001) Sanitation on Ships: Compendium of Outbreaks of Foodborne and Waterborne Disease and Legionnaires' Disease Associated with Ships 1970-2000.

8. McCarter, Y. (2009) Infectious Disease Outbreaks on Cruise Ships. Clinical Microbiology Newsletter, 31, 161-168.

9. Mouchtouri, V., Bartlett, C., Diskin, A. and Hadjichristodoulou, C. (2012) Water Safety Plan on Cruise Ships: A Promising Tool to Prevent Waterborne Diseases. Science of the Total Environment, 429, 199-205.

10. Waterman, S., Demarcus, T., Wells, J. and Blake, P. (1987) Staphylococcal Food Poisoning on a Cruise Ship. Epidemiology and Infection, 99, 349-353.

11. Azara, A., Piana, A., Sotgiu, G., Dettori, M., Deriu, M., Masia, M., Are, B. and Muresu, E. (2006) Prevalence Study of Legionella spp. Contamination in Ferries and Cruise Ships. BMC Public Health, 6, 1-6.

12. Dahl, E. (2006) Norovirus Challenges Aboard Cruise Ships. International Maritime Health, 57, 230-234.

13. Marucheck, A., Greis, N., Mena, C. and Cai, L. (2007) Product Safety and Security in the Global Supply Chain: Issues, Challenges and Research Opportunities. Journal of Operations Management, 29, 707-720.

14. Said, B., Ijaz, S., Kafatos, G., Booth, L., Thomas, L., Walsh, A., Ramsay, M. and Morgan, D. (2009) Hepatitis E Outbreak on Cruise Ship. Emerging Infectious Diseases, 15, 1738-1744.

15. Weaver, A. (2005) The Mcdonaldization Thesis and Cruise Tourism. Annals of Tourism Research, 32, 346-366.

16. The World Tourism Organization (2010) Cruise Tourism: Current Situation and Trends. Madrid, UNWTO.

17. Folinas, D., Manikas, I. and Manos, B. (2006) Traceability Data Management for Food Chains. British Food Journal, 108, 622-633.

18. Regattieri, A., Gamberi, M. and Manzini, R. (2007) Traceability of Food Products: General Framework and Experimental Evidence. Journal of Food Engineering, 81, 347-356.

19. Wu, L., Xu, L. and Gao, J. (2011) The Acceptability of Certified Traceable Food among Chinese Consumers. British Food Journal, 113, 519-534.

20. Mai, N., Bogason, S., Arason, S., árnason, S. and Matthiasson, T. (2010) Benefits of Traceability in Fish Supply Chains-Case Studies. British Food Journal, 112, 976-1002.

21. Engelseth, P. (2009) Food Product Traceability and Supply Network Integration. Journal of Business & Industrial Marketing, 24, 421-430.

22. Sperber, W. (2005) HACCP Does Not Work from Farm to Table. Food Control, 16, 511-514.

23. Matsuda, T. (2007) Gap as a Baseline, Traceability as a Pipeline to Build Consumer Confidence. Food and Fertilizer Technology Center, 5-15.

24. The Food Marketing Research and Information Center (FMRIC) (2008) Handbook for Introduction of Food Traceability Systems (Guidelines for Food Traceability). 2nd Edition, FMRIC, Tokyo, 1.

25. Smith, G., Tatum, J., Belk, K., Scanga, J., Grandin, T. and Sofos, J. (2005) Traceability from a US Perspective. Meat Science, 71, 174-193.

26. Golan, E., Krissoff, B. and Kuchler, F. (2004) Food Traceability: One Ingredient in a Safe and Efficient Food Supply. USDA, Economic Research Services, 2, 19.

27. The Food Standard Agency (FSA) (2002) Traceability in the Food Chain. 19-47.

28. Stringer, M. and Hall, M. (2007) A Generic Model of the Integrated Food Supply Chain to Aid the Investigation of Food Safety Breakdowns. Food Control, 18, 755-765.

29. Chrysochou, P., Chryssochoidis, G. and Kehagia, O. (2009) Traceability Information Carriers, the Technology Backgrounds and Consumers' Perceptions of the Technological Solutions. Appetite, 53, 322-331.

30. Xiong, B., Fu, R., Lin, Z., Luo, Q., Yang, L. and Pan, J. (2010) A Solution on Pork Quality Traceability from Farm to Dinner Table in Tianjin City, China. Agricultural Sciences in China, 9, 147-156.

31. Senneset, G., Forås, E. and Fremme, K. (2007) Challenges Regarding Implementation of Electronic Chain Traceability. British Food Journal, 109, 805-818.

32. Sauvage, T. (2003) The Relationship between Technology and Logistics Third-Party Providers. International Journal of Physical Distribution and Logistics Management, 33, 236-253.

33. Lee, D. and Park, J. (2008) RFID-Based Traceability in the Supply Chain. Industrial Management and Data Systems, 108, 713-725.

34. Kros, J., Richey, R., Chen, H. and Nadler, S. (2011) Technology Emergence between Mandate and Acceptance: An Exploratory Examination of RFID. International Journal of Physical Distribution and Logistics Management, 41, 697-716.

35. Kück, M. (2006) Chill-On: Ideas and Solutions to Improve Food Quality, Safety and Traceability.

36. Finn, M., Walton, M. and Elliott-White, M. (2000) Tourism and Leisure Research Methods. Pearson Education, Harlow, 81.

37. Miao, M. (2010) Critical Success Factors for Implementing Traceability Systems in Chinese Food Enterprises. Ph.D. Dissertation, University of Bedfordshire, Bedford, 99-101.

38. Zhang, X., Zhang, J., Liu, F., Fu, Z. and Mu, W. (2010) Strengths and Limitations on the Operating Mechanisms of Traceability System in Agro Food, China. Food Control, 21, 825-829.

39. McMeekin, T., Baranyi, J., Bowman, J., Dalgaard, P., Kirk, M., Ross, T., Schmid, S. and Zwietering, M. (2006) Information Systems in Food Safety Management. International Journal of Food Microbiology, 112, 181-194.

40. Poghosyan, A., Gonzalez-Diaz, F. and Bolotova, Y. (2004) Traceability and Assurance Protocols in the Global Food System. International Food and Agribusiness Management Review, 7, 118-26.

41. Bechini, A., Cimino, M., Lazzerini, B., Marcelloni, F. and Tomasi, A. (2005) A General Framework for Food Traceability. Proceedings of the Symposium on Applications and the Internet Workshop, Trento, 31 January-4 February 2005.

42. van Rijswijk, W., Frewer, L.J., Menozzi, D. and Faioli, G. (2008) Consumer Perceptions of Traceability: A Cross-National Comparison of the Associated Benefits. Food Quality and Preference, 19, 452-464.

13
CHAPTER

INTERNATIONAL CONSUMER INSIGHTS INTO THE DESIRES AND BARRIERS OF DINERS IN CHOOSING HEALTHY RESTAURANT MEALS

R.S. Newson[a], R. van der Maas[a], A. Beijersbergen[a], L. Carlson[b], C. Rosenbloom[c]

[a]*Research and Development, Unilever, Vlaardingen, The Netherlands*

[b]*Research and Development, Unilever, Englewood Cliffs, USA*

[c]*Division of Nutrition, Georgia State University, USA*

ABSTRACT

Overweight and obesity are global public-health problems and unhealthy restaurant meals have been identified as one contributing factor. Given the increase in restaurant meals and the number of chefs and restaurants throughout the world, small changes in restaurant meals can have a large public health impact. However, to ensure that chefs and operators are able to provide changes that diners will accept and find appealing, an understanding of diners' desire for healthier menus items and the barriers faced in choosing healthier meals is required. As such we conducted an international consumer study to identify these barriers and needs. A cohort of restaurant diners was recruited from ten countries: United Kingdom, United States of America, Germany, Poland, Turkey, Russia, Brazil, South Africa, Indonesia and China ($n = 5000$, aged 18–65 years, 50% females). Participants completed a comprehensive web-based questionnaire on aspects related to healthy eating at restaurants. Globally only 18% of the sampled diners were strongly satisfied with current healthy options on restaurant menus. Among the diners there was a preference to have "slightly healthier" options. The top 3 small changes that these diners wanted to see included on restaurant menus to make them healthier included: steamed, baked or grilled instead of fried, fresh ingredients used, and served with plenty of vegetables. Taste, price and satiation were seen as key barriers to current healthy options. Diners had clear preferences for when they wanted to see healthy items on the menu in terms of time of day, time of the week and occasions. Country, age and gender had a large influence on preferences, while personal factors such as diet type, family status and food reactions had minimal influence. In the current study, diners were largely unsatisfied with current healthy options and clear barriers and triggers for healthier meals were identified. Nutritionists, dietitian, chefs and managers working for restaurants and other out of home food services can leverage these insights to provide healthier and appealing meals.

Keywords: Food service; Nutritious meals; Public health; Behavior change; Restaurant meals

1. INTRODUCTION

Since 1980 rates of obesity have doubled globally, in 2008 more than 1.4 billion adults aged 20 and over were overweight and 500 million of these were obese (World Health Organisation, 2014). Overweight and obesity are a central public health issue as they are known to increase the risk of morbidity and mortality from diabetes, cardiovascular disease, musculoskeletal disease and some cancers (American Heart Association, 2013). These diseases impact quality of life and contribute to burgeoning health care costs. For example, in the USA the national cost of overweight and obesity combined has been estimated at $148.9 billion (Gilden Tsai, Williamson, & Glick, 2011). Overweight and obesity are largely preventable through lifestyle modifications, including calorie reduction.

Food eaten in restaurants has a significant impact on caloric intake and therefore weight gain. One study showed that each additional meal or snack eaten away from home adds an average of 134 calories that day, compared with the same meals or snacks prepared at home (Todd, Mancino, & Lin, 2010). Holding all else constant, one additional meal eaten away from home each week could result in about two extra pounds per year. The effect of food prepared away from home on daily caloric intake is even more pronounced in obese individuals. An away-from-home meal adds an average of 239 calories to daily caloric intake for obese individuals (Body Mass Index (BMI) greater than or equal to 30 kg/m^2) versus 88 additional calories for those with a BMI less than 25 kg/m^2 (McCrory et al., 1999). It is therefore not surprising that an increased frequency of eating away from home has been associated with long-term weight gain (McCrory et al., 1999).

In the foodservice industry there are 17.5 million outlets around the world with more than 500 billion transactions daily (Data Monitor, 2013). With tens of millions of chefs and cooks in the world preparing meals and each one cooking hundreds of thousands of meals in their lifetime, it is clear that the food service industry, and associated partners, have an opportunity to impact on the global obesity epidemic.

Small changes can make a large difference at the population level (e.g., reducing obesity) and over time many small changes in dishes could also make a difference at the individual level (e.g., helping individuals to meet daily nutritional recommendations). Given the large number of chefs and operators in the world, making small changes in their top dishes could really add up to make a big difference toward energy intake and the overweight and obesity epidemic. Moreover, behavioral economists suggest that "nudging" people to make small changes is the key to treating obesity as these changes are achievable and sustainable as well as promoting self-interest (such as choosing healthier foods) without taking away freedom of choice (Heshmat, 2011 and Thaler and Sunstein, 2013). Additionally, enabling environments play an important role in overeating, therefore helping to provide a supportive environment to help individuals make it easier to make healthy choices without relying on individual "will power" could be a realistic public health approach applied to the food service industry (Wansink, Just, & Payne, 2009).

The need for changes to current restaurant options to improve the level of healthiness is clear for diners. However, to ensure that chefs and operators are able to provide changes that diners will accept and find appealing, an understanding of diner's desire for healthier menus items and the barriers faced in choosing healthier meals is required. To explore this, we conducted an international online cohort study in 10 countries to understand current levels of satisfaction with healthy options, evaluate what diners wanted to see from healthy meals (extent of change toward healthiness, what items should be included to make menu's healthy and when they want healthier meals) and to evaluate the potential barriers of three important facets of restaurant meals: price, taste and satiation.

2. MATERIALS AND METHODS

2.1. Design and procedure

The study consisted of an online cohort conducted in the following 10 countries: United Kingdom, United States of America, Germany, Poland, Turkey, Russia, Brazil, South Africa, Indonesia and China. These countries were selected to provide a cross-section of different cultures, economies and cuisines. Data was collected during October to November 2011 using an online questionnaire, received via an email link, which took approximately 30-min to complete. This was completed using the participants own computer in their own environment. The questionnaire had to be completed in one sitting and prompting was provided on incomplete answers to reduce the occurrence of missing data. Participants were recruited through an international online panel provider (BrainJuicer), who was registered with the Data Protection Act. Informed consent was provided by all participants and they received a minor reimbursement for their time.

2.2. Participants

A recruitment target was set for each country of a sample size of 500 adults aged 18–64 years with an equal representation of males and females (total sample, $n = 5000$). The online participant panels were stratified based on gender to enable cross-country comparability. To ensure the relevant audience for the questionnaire, the following inclusion criteria were applied: must eat a meal out of home at least once a week (e.g., at a restaurant) and must not be employed in the hospitality or marketing industry.

2.3. Questionnaire

A self report questionnaire was developed in consultation with experts in food services, nutrition and consumer science. Questions were developed, based on common themes seen in consumer and food science research, to elicit information to understand diner's reactions to meals eaten away from home, their desires when dining out, and perceived barriers faced in choosing healthier meals. The term "healthy" was not defined to respondents in the study as we wanted to learn without the complications of definitions what current diner's wanted and what their barriers were. One standard framework questionnaire was designed, which was then adapted to each country for local foods, serving sizes and culture by local chefs and nutritionists. When a language other than English was used, this was then translated into the local language by a professional service.

2.3.1. Socio-demographics and personal characteristics

Participants were asked about several personal factors which may relate to eating at a restaurant. Participants reported whether they had children living at home, whether they had food allergies or intolerances and their main dietary preference: meat eater/vegetarian/vegan/pescatorian (only eats fish as a source of protein). Finally, to understand the restaurant eating behavior of diners, the participants were asked how often they chose healthy menu items at restaurants and how often they ate at restaurants. First, participants reported how often they chose a healthy option from a restaurant menu on a 4-point scale (not at all often/not very often/quite often/very often). Then participants reported how often they ate at restaurants for breakfast, lunch and dinner: every day/6 times a week/5 times a week/4 times a week/3 times a week/2 times a week/once a week/once every 2 weeks/once a month/less often and never. These were converted per meal occasion into times per week to create a continuous variable (e.g., once a month = .25 times per week etc.). These three meal occasions were then summed to create total meals eaten in restaurants per week.

2.3.2. Current satisfaction with healthier restaurant meals

To determine the current level of satisfaction that diner's experience with currently offered healthy options, participants were asked to what extent they agreed with the following healthy options: "I am happy with the healthy options available when eating out". Participants could answer on a 5-point scale (strongly disagree/slightly disagree/neutral/slightly agree/strongly agree).

2.3.3. The needs of diners in choosing healthier restaurant meals

We asked the participants how much change they wanted, what they wanted changed, and when they wanted it. Participants were asked to rate their level of interest in seeing small healthy changes in out-of-home meals by reporting the extent to which they agreed with the following statement: "I would prefer to have slightly healthier food options when eating out." Participants could answer on a 5-point scale (strongly disagree/slightly disagree/neutral/slightly agree/strongly agree).

To establish which food options diners wanted to see in restaurants in order to make current restaurant menu's healthier, participants were shown a list of 20 items (presentation randomized to prevent bias) and asked to select as many as applied. The items consisted of:

Preparation techniques:

- Steamed, baked or grilled instead of fried.
- Prepared with healthy oils/fats.
- Locally grown.
- Produced ingredients.

Adding healthy ingredients:

- Plenty of vegetables.
- Fish dishes.
- Lean meat.
- Grilled vegetables.

Removing 'unhealthy' ingredients:

- Low in fat.
- Without additives.
- Lower in calories.
- Low in sugar.
- Gluten free.
- No cream based sauces.
- Not seasoned with salt.

Serving style:

- Right-sized portions.
- Smaller or half sized portions.
- *Offering alternative options:*
- Healthier side dishes.
- Whole grain options.
- Vegetarian dishes.

To establish when diners are most interested in seeing healthy items on the menu, they were asked to rate on a 4-point scale (not at all often/not very often/quite often/very often) how likely they were to choose a healthy item on the following occasions:

Time of day:

- Breakfast.
- Lunch.
- Dinner.

Time of the week:

- Monday.
- Tuesday.
- Wednesday.
- Thursday.
- Friday.
- Saturday.
- Sunday.

Events/purposes:

- When I am not feeling 100% healthy.
- When life is hectic.
- When someone else with me chooses a healthy option.
- When out on a business meeting.
- When I am trying to manage my weight.
- Depending on the weather.
- In preparation for a big event (e.g., wedding or holiday).

2.3.4. The barriers diners face in choosing healthy restaurant meals

Price, taste and expected satiation are important in getting diners to select a dish from a menu; as such it is important to understand how diners currently rate healthy options in terms of these three facets. We asked diners to tell us how much they agreed with the following three statements: "Healthy options tend to be more expensive", "Healthy options on the menu usually don't sound very tasty" and "Often the healthy option is not very filling." Participants could rate each statement on a 5-point scale (strongly disagree/slightly disagree/neutral/slightly agree/strongly agree).

2.4. Analyses

The majority of results are presented as descriptive statistics, using percentages or means and standard deviations where relevant. To test for differences based on socio-demographic or personal characteristics Chi-square analyses were used for categorical outcomes and Analysis of Variances (ANOVAs) were used for continuous outcomes. Due to the large number of countries included, it was decided for clarity of reporting to present all tables and figures globally, as well as for individual countries. All analyses were conducted using SPSS version 18.0 and significant differences were evaluated at a level of $p < .05$.

3. RESULTS

Characteristics of the participants are displayed in Table 1, for the total sample and per country. More than half of the participants globally had children under the age of 18 at home (58.4%), ranging from 41.0% in the USA to 74% in China. Notably, almost one fifth of the population self-reported experiencing a food allergy or intolerance; this was lowest in the UK and highest in Indonesia where one quarter of participants reported an issue with food. Ninety percent of the global population were meat eaters, China and Indonesia had the highest proportion of non-meat eaters, with 13% and 15% of the populations respectively only eating fish as a main protein source.

Interestingly, 66% of the population self-reported that they "very often" or "quite often" chose healthy meals. This was higher in the (South-East) Asian regions, and lower in the European regions. On average globally, these participants ate at restaurants seven times per week; which was expected as this sample was selected to ensure it focused on people that ate at restaurants. China had the highest rate of restaurant eaters, recording on average 11 restaurant meals per week and the UK recorded the lowest. Globally, lunch was the most popular meal eaten at restaurants, where on average this was 3.3 times per week.

Table 1. Background characteristics of participants ($n = 5000$; 500 per country).

	Global (%)	USA (%)	UK (%)	Poland (%)	Brazil (%)	China (%)	S. Africa (%)	Germany (%)	Russia (%)	Turkey (%)	Indonesia (%)
Gender											
Females	50.0	50.4	50.2	50.8	50.0	50.0	50.0	49.6	50.0	49.2	50.0
Age groups											
18–24 years	14.8	5.6	6.4	20.6	27.8	11.2	14.2	11.0	15.6	13.6	22.0
24–34 years	36.6	17.4	28.8	49.6	37.4	53.6	28.8	30.4	44.0	43.2	32.6
35–44 years	24.1	15.8	23.8	17.4	21.4	27.6	25.4	26.6	24.0	33.8	25.6
45–54 years	16.0	26.2	25.0	9.2	10.2	6.6	22.2	22.8	13.8	8.0	16.2
55–64 years	8.5	35.0	16.0	3.2	3.2	1.0	9.4	9.2	2.6	1.4	3.6
Family situation											
Children at home	58.4	41.0	47.8	62.4	56.2	74.0	58.4	45.6	70.2	63.8	64.8
Food reactions											
Allergy/intolerance	17.1	14.0	7.8	15.6	19.4	22.6	24.4	12.0	16.6	13.8	25.2
Main dietary type											
Vegetarian	2.7	2.4	4.0	1.0	3.4	3.2	1.4	4.2	1.6	.6	5.0
Vegan	2.0	1.2	1.2	.2	6.2	3.0	1.0	1.4	1.6	1.8	2.0
Pescatorian[a]	5.7	2.0	2.4	3.4	4.8	13.4	4.4	3.8	4.0	4.2	14.4
Meat eater	89.7	94.4	92.4	95.4	85.6	80.4	93.2	90.6	92.8	93.4	78.6
Frequency of choosing healthy											
Very or quite often[b]	65.6	54.6	58.8	50.4	74.2	87.6	68.6	59.6	52.6	68.0	81.8
	M ± SD	M ± SD	M ± SD	M ± SD	M ± SD	M ± SD	M ± SD	M ± SD	M ± SD	M ± SD	M ± SD
Restaurant meals per week[c]											
Breakfast	2.1 ± 2.3	1.4 ± 2.0	1.4 ± 2.0	2.2 ± 2.3	2.4 ± 2.6	4.3 ± 2.3	1.7 ± 2.2	1.7 ± 2.0	1.7 ± 2.2	2.0 ± 2.2	2.1 ± 2.3
Lunch	3.3 ± 2.2	2.3 ± 2.0	2.6 ± 2.0	2.3 ± 1.8	3.7 ± 2.4	4.2 ± 2.0	2.6 ± 2.1	2.8 ± 2.1	4.2 ± 2.0	3.5 ± 2.3	4.3 ± 2.2
Dinner	1.7 ± 1.7	2.0 ± 1.8	1.4 ± 1.7	1.3 ± 1.4	1.9 ± 1.8	2.5 ± 1.9	1.7 ± 1.8	1.3 ± 1.5	1.4 ± 1.5	1.5 ± 1.4	2.3 ± 2.1
Total meals	7.1 ± 4.8	5.7 ± 4.6	5.4 ± 4.6	5.9 ± 4.0	8.0 ± 5.0	11.0 ± 4.6	5.9 ± 4.6	5.9 ± 4.3	7.3 ± 4.1	7.0 ± 4.3	8.6 ± 4.8

[a]A pescatorian eats only fish as the meat source.

[b]This represents participants that reported they" very often" or "quite often" chose a healthy meal when eating out at restaurants.

[c]This is the average number of meals eaten out per week per individual meal occasion, and summed for 'total meals'.

3.1. Current satisfaction with healthier restaurant meals

The level of satisfaction of diners, both globally and for individual countries, is shown inFig. 1. Globally, only 18% of diners were completely satisfied with the current healthy options on restaurant menus, with 43% noting room for improvement. Chi-square analyses of independence showed that there were significant differences between the countries in terms of level of satisfaction ($\chi^2 = 414.9$, $p < .001$). Level of satisfaction was notably highest in China and Indonesia, where 46% and 32% of diners were completely satisfied with current healthy options. In the USA and Germany this fell to only 10% of diners. Men were significantly more satisfied than women with current healthy options ($\chi^2 = 4.16$, $p < .05$) while age had no influence on level of satisfaction ($p > .05$).

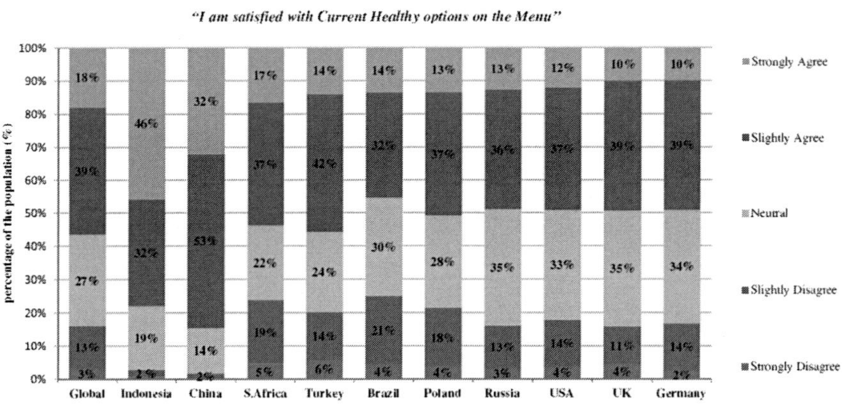

Figure 1. The level of satisfaction of guests with current healthy options on the menu.

3.2. The needs of diners in choosing healthier restaurant meals

The reaction of the diners regarding whether they would prefer to have slightly healthier food options when eating out is presented in Fig. 2, the outcome is treated as continuous to provide a mean rating. Both globally, and across all countries, participants were on average above neutral in terms of their interest in the idea of small, healthier changes to current restaurant menus. The global average was just below "slightly agree." The difference between countries was significant ($F(1,9) = 22.9$, $p < .001$). China and Brazil had the highest interest among all countries. Women were significantly more interested in small changes than men, ($F(1,1) = 31.8$, $p < .001$). There was also a significant influence of age group with diners aged 35–44 being the most interested in small changes

($F(1,4) = 8.2, p < .001$). However, all groups were interested in small changes but, these analyses just highlight that some groups were relatively more interested than others.

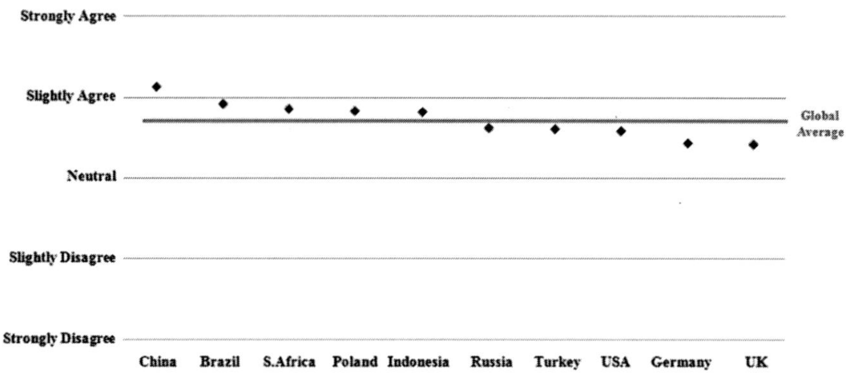

Figure 2. The reaction of guests to the notion of small changes in restaurant menu's.

To make useful changes in current menus it is important to know what diners want included on restaurant menus to make the offerings healthier and appealing. This will reflect a combination of what diners like, what they believe is currently missing from restaurant menus, and what they think defines healthy in a dish. Table 2 shows the items that diners wanted to see in restaurants. Diners were most interested in seeing changes in preparation techniques. The highest ranking items globally were 'steamed, baked or grilled instead of fried' (64.0%) and fresh ingredients used (63.6%). Plenty of vegetables, raw or cooked (59.0%) and items low in fat (58.8%) were also popular. There were significant differences in selections based on country (χ^2 on top 3 items ranged from 95.6–218.4, $p < .001$). 'Without additives' was high in the rankings for Turkey, China and Indonesia. Of the 20 changes listed in Table 2, globally, diners indicated that they wanted to see on average 8.9 items out of the 20 choices available on restaurant menu's (SD = 4.8). This ranged from 7.5 items in China (SD = 4.2) to 10.4 items in South Africa (SD = 4.7). Only 1% of respondents did not select any of the presented items. Women were more interested than men in seeing steamed, baked or grilled items, plenty of vegetables and fresh ingredients in restaurants (χ^2 ranged from 36.2–13.3, $p < .001$). Older adults were significantly more interested in seeing steamed, baked or grilled items and plenty of vegetables on the menu, compared to younger adults (χ^2 ranged from 40.3 to 62.9, $p < .001$), but there was no age difference on wanting to see fresh ingredients used ($p > .05$). Note, the data split by gender and age groupings are not shown.

Table 2. Items that diners wanted to see in restaurants to make healthier menu's (n = 5000; 500 per country).

	Global (%)	USA (%)	UK (%)	Poland (%)	Brazil (%)	China (%)	S.Africa (%)	Germany (%)	Russia (%)	Turkey (%)	Indonesia (%)
Preparation techniques											
Steamed, baked or grilled	64.0	70.6	61.6	71.0	70.2	48.8	77.8	44.6	64.0	68.6	58.6
Fresh ingredients used	63.6	47.0	50.4	64.8	55.8	61.2	67.6	65.2	63.6	68.8	81.0
Prepared with healthy oils/fats	54.9	56.4	56.2	51.6	49.0	57.8	60.8	51.2	54.9	67.8	58.4
Local ingredients	31.6	31.6	32.2	24.2	28.0	22.6	32.4	32.4	31.6	46.2	34.0
Adding healthy ingredients											
Plenty of vegetables	59.0	59.8	55.8	65.4	55.6	45.4	68.8	59.6	59.0	51.0	64.4
Fish dishes	55.0	44.6	39.8	61.8	58.2	54.6	57.8	52.0	55.0	66.8	49.0
With lean meat	52.3	56.6	48.4	60.6	59.8	31.2	62.4	49.6	52.3	47.2	59.2
Grilled vegetables	41.3	58.0	40.4	41.0	48.6	13.2	58.6	38.4	41.3	52.4	15.2
Removing 'unhealthy' ingredients											
Low in fat	58.8	56.2	55.8	53.8	67.0	53.4	63.4	51.6	58.8	67.0	74.6
Without additives	52.1	43.8	40.0	19.8	41.6	61.2	50.6	46.6	52.1	77.4	73.2
Lower in calories	43.5	48.6	48.0	38.2	45.8	45.8	49.2	33.0	43.5	44.2	50.8
Low in sugar	41.0	42.4	45.2	28.6	45.0	41.0	53.6	35.2	41.0	36.6	55.2
Gluten free	27.2	23.2	16.8	16.0	28.2	13.6	29.6	23.6	27.2	35.4	37.6
No cream based sauces	22.3	19.8	27.2	32.2	20.2	18.4	29.2	21.0	22.3	15.6	19.2
Not seasoned with salt	22.1	38.2	33.4	30.2	21.2	13.4	25.4	10.6	22.1	30.6	7.8
Serving style											
Right sized portions	42.3	52.2	48.2	30.2	40.4	34.8	49.2	33.4	42.3	34.0	62.6
Smaller or half sized portions	27.4	41.6	35.0	23.2	36.8	13.2	42.4	28.8	27.4	20.2	20.0
Offering alternative options											
Healthier side dishes	57.1	60.0	50.8	54.0	55.2	60.2	70.8	54.0	57.1	44.6	63.0
Wholegrain options	39.4	50.6	34.2	49.2	49.0	15.8	58.2	40.8	39.4	41.4	27.0
Vegetarian dishes	32.0	30.2	31.0	26.4	34.0	46.8	37.0	32.4	32.0	12.6	40.6

For chefs and operators to provide the most appealing menus it is important to know when they should put most emphasis on healthy menu items versus occasions when diners just want to treat themselves. As such we explored how likely diners were to choose a healthy option from the menu on several different occasions. The occasions when diners were likely to choose healthy options are presented in Table 3 by time of day, time of week, and purpose. All country differences on all options were significant at$p < .05$, indicating large cultural differences between the evaluated countries. In terms of time of day, lunch was globally reported as the time of day when diners are most likely to choose a healthy option (65.4%), followed by dinner and then breakfast. This was largely due to lunch being the most common occasion to eat at a restaurant. At an individual country level, China and Russia rated breakfast as the most likely time to eat healthily, and USA, Poland, Turkey and Indonesia rated dinner as the most likely time. For time of the week, on average diners were more likely to choose healthy items on weekdays (Monday–Friday) and least likely to choose healthy items on Saturday and Sunday. This trend was repeated in all individual countries with the exception of Brazil, where diners were more likely to choose healthy on a Friday and Saturday night, compared to the rest of the days of the week. Purpose of healthy eating was relatively consistent across most countries, although the ratings varied in level, with 'trying to manage weight', 'when I am not feeling healthy' and 'when someone else chooses healthy' being the top three reported reasons for choosing healthy. The only exceptions to this were people in the United Kingdom, who also rated 'getting ready for an event' in the top three, and Brazil who reported 'when life is hectic' in the top three reasons to choose healthy. The majority of age and gender differences were significant at $p < .05$, in general women selected all of the "healthy eating occasions" options more frequently with men and the selections decreased with increasing age.

Table 3. Factors influencing healthy choices on menu items (n = 5000; 500 per country).

	Global (%)	USA (%)	UK (%)	Poland (%)	Brazil (%)	China (%)	S.Africa (%)	Germany (%)	Russia (%)	Turkey (%)	Indones (%)
Occasions for healthy eating[b]											
Time of day											
Breakfast	58.9	35.8	45.4	65.4	70.2	77.2	57.6	46.0	62.4	64.2	64.4
Lunch	65.4	56.0	66.8	58.2	83.0	72.8	72.4	56.8	54.6	62.8	70.2
Dinner	61.7	57.0	55.4	58.8	70.6	74.4	63.6	52.2	47.4	67.4	70.4
Time of week											
Monday	62.0	54.4	60.2	59.6	69.4	74.0	69.0	45.4	60.2	61.2	67.0
Tuesday	61.8	54.2	60.2	57.8	69.6	72.8	70.6	45.6	58.8	60.8	68.0
Wednesday	62.1	53.8	60.8	57.6	68.4	76.4	68.2	46.6	58.2	61.6	69.4
Thursday	61.4	54.0	60.0	55.6	68.0	75.4	67.6	46.8	57.6	60.6	68.0
Friday	59.1	49.0	53.6	55.6	75.4	74.4	54.4	49.0	51.4	59.8	68.2
Saturday	53.3	45.4	46.2	50.8	73.2	71.4	45.6	46.2	41.8	50.8	62.0
Sunday	53.0	46.2	47.6	51.2	68.4	70.0	52.0	45.6	40.8	49.8	58.0
Purpose											
Trying to manage weight	68.7	67.0	71.2	71.8	66.6	69.6	79.0	64.4	59.8	67.6	70.2
When I am not feeling healthy	63.8	54.2	60.2	70.8	63.0	72.4	67.0	52.6	61.4	66.4	70.2
Someone else chooses healthy	61.1	52.0	53.4	65.0	64.6	82.8	64.4	47.4	52.0	62.4	67.4
At a business meeting	52.4	44.6	46.6	57.8	61.6	59.2	60.2	45.4	47.4	54.0	47.0
In preparation for an event	49.4	41.4	55.0	56.4	62.8	58.8	51.8	39.6	38.8	47.6	41.8
When life is hectic	45.5	32.0	38.8	47.2	68.4	57.0	42.4	37.4	48.4	43.6	39.4
Depending on the weather	43.1	31.4	49.0	43.4	49.0	55.2	43.4	38.4	32.6	43.8	45.2
Barriers to healthy eating[a]											
Healthy options are expensive	57.1	56.2	45.6	69.4	59.2	65.8	57.4	48.6	51.4	60.2	57.0
Healthy options are not tasty	43.1	40.2	46.6	41.6	38.8	38.0	43.4	31.6	58.0	46.2	46.8
Healthy options are not filling	45.4	41.6	48.8	51.4	39.2	41.6	47.2	27.8	63.8	47.2	45.0

Bold items indicate the top 3 items per category.

[a] Percentages reflect those participants who "slightly" or "strongly" agreed to these statements.

[b] Percentages reflect those participants that recorded that they were likely to choose healthy on these occasions: "very often" or "quite often".

3.3. The barriers diners face in choosing healthier restaurant meals

Developing menus to ensure that diners want to order the healthier items requires an understanding of what the diners think of current healthy options. Table 3 presents the diner reactions to perceived barriers to choosing healthy options on menus. Globally, 57.1% of diners reported that they thought that healthy items tended to be more expensive, 43.1% of diners thought that healthy options don't usually sound as if they would be tasty and 45.4% of diners conveyed that often healthy options are not filling. There were significant differences between the countries in terms of taste ($\chi^2 = 90.7, p < .001$), satiation ($\chi^2 = 155.3, p < .001$) and expense ($\chi^2 = 97.7, p < .001$). Russian diners were more convinced than other countries that healthy items were not very tasty or satiating. Among all countries Brazil had the highest opinion of healthy menu items, having among the lowest ratings of dissatisfaction with the taste and satiation of healthy dishes. Compared to women, men thought that healthier dishes were less tasty (47% compared to 43%, $\chi^2 = 27.9, p < .001$), and less satiating (51% compared to 45%, $\chi^2 = 54.9, p < .05$). There was no difference in terms of expense ($p > .05$). Compared to older adults, younger adults thought healthy meals were less tasty (46% in adults aged 18–24 compared to 37% in adults aged 55–64: $\chi^2 = 23.7, p < .001$), less satiating (47% in adults aged 18–24 compared to 37% in adults aged 55–64: $\chi^2 = 22.7, p < .001$), and more expensive (59% in adults aged 18–24 compared to 50% in adults aged 55–64: $\chi^2 = 30.1, p < .001$). Correlational analyses were used to evaluate the associations between ratings on expense, taste and satiation of current healthy menu offerings. Analyses revealed that ratings on taste and satiation were strongly positively correlated to each other ($r = .528, p < .001$), indicating that if a diner thought a healthy item was less tasty, they also thought it was less satiating. The association of taste and satiation to expense was also positive and significant, but the associations were less strong (association between taste and expense, $r = .290, p < .001$ & satiation to expense: $r = .352, p < .001$).

3.4. Diner reactions in different target groups

When looking at those who identified as non-meat eaters, 30% of vegetarians/vegans/pescatorians were satisfied with current healthy options, which was significantly higher than meat-eaters (only 16.6% were satisfied: $\chi^2 = 58.57, p < .001$). They were also more interested in small changes than those that chose a meat-based diet ($F(1,1) = 79.2, p < .001$).

Diners who reported food allergies or intolerances were more satisfied with current healthy options (21%) than those without food allergies or intolerances (17.2%: $\chi^2 = 9.1, p < .001$) and they were more interested in small changes ($F(1, 1) = 13.4, p < .01$). More people with food allergies or intolerances reported that healthy meals were expensive (60.3% vs. 56.4% among people without food allergies of intolerances: $\chi^2 = 4.5, p < .05$).

4. DISCUSSION

The current study sought to provide new international insights into the desires and barriers of diners in choosing healthy restaurant meals. In light of this we evaluated what diners wanted to see on restaurant menu's, what changes they wanted to see made, when they wanted healthier meals and understand the importance of the potential barriers of price, taste and satiation.

Remarkably, only 18% of diners globally surveyed were completely satisfied with current healthy options on restaurant menus. Notably, 43% indicated room for improvement. This suggests a definite need to change for chefs and operators in their current menus and overall should act as a wakeup call to the industry. This is particularly important as higher satisfaction with current healthy options was associated with eating at restaurants more often and every restaurant wants to provide satisfaction to its patrons and potentially increase the number of diners.

The majority of diners were interested in seeing small, healthier changes to the menu. This supports the notion that small changes are an important strategy to making improvements to the healthiness of restaurant menus. Diners are caught at the intersection of wanting to eat healthy but also wanting to treat themselves on the weekend or for special occasions. Small changes can appeal to a diner whereas a total overhaul of a restaurant menu or favorite dish is less appealing to diners. A joint report of the American Society for Nutrition, the Institute of Food Technologists, and the International Food Information Council advocates for the small change approach to help address the obesity epidemic (Hill, 2009). The report identifies several ways that small changes, such as the ones identified by the participants of our research, could have a positive impact on health. This is important as small changes are more feasible to achieve and maintain than large changes and that even small changes can have a big effect on body weight regulation. Furthermore, the small change approach can help mitigate the environmental factors that encourage and promote overeating, which can work for or against those who want to reduce calories (Wansink et al., 2009). We tend to underestimate calories consumed in restaurant meals and we eat "mindlessly." By making small changes to restaurant meals, chefs and food service operators could make a valuable contribution to reducing energy intake and increasing healthiness of food consumption of diner's throughout the world. Moreover, given the high level of interest of diners in seeing more items on the menu which are nutrient dense, such as more vegetables and fish dishes, small changes made by chefs may also be able to impact on the overall energy density of diner's diets which is linked to higher satiation and better health (Nicklas, Drewnowski, & O'Neil, 2014).

Diners in our study wanted to see changes made to preparation techniques with the most interest in steaming, baking, and grilling foods, along with the use of fresh ingredients prepared using healthier fats and oils. They also wanted to see the addition of healthy ingredients including the use of more vegetables, fish, and lean cuts of meat. These changes point to not only what diners want but what they see missing in today's restaurant offerings. It is also interesting that the positive changes were more often mentioned than negative changes (i.e., removing ingredients deemed to be unhealthy such as cream-based sauces or high in sodium). This implies that chefs and food service operators have an opportunity to add healthy ingredients and promote healthy food preparation techniques without dwelling on the negatives (low fat or low sodium). There were strong gender, age and country influences on the small changes that diners wanted to see, which should be taken into account by chefs and food service operators in various countries.

Understanding when diners want healthier meals could help chefs and operators to decide when they need to increase or highlight healthier options. Certain meals (lunch, dinner), days of the week (Monday to Friday) and certain occasions (managing weight, when a friend chooses healthy or when not feeling well) were key triggers for choosing healthy options. This is of note to chefs and operators, as they know that most diners on these occasions are already interested in eating more healthily, but it indicates that extra effort needs to be made on other occasions (e.g., breakfast and on weekends when diners are not as likely to be looking for healthy options). Interestingly choosing these options were strongly related to frequency of eating out, indicating that people who eat out more frequently are more likely to eat more healthily than those who eat out less. It is therefore important for chefs and operators to be aware that they may have different challenges from "regular" diners from those who are less frequent diners. In a study on healthy dining, it was found that subtle reminders on health and nutrition on the menu increased the selection of healthier options among dieters (Papies & Veling, 2013). It was concluded that expanding the options of tasty, less caloric foods at comparable prices to other offerings of the menu could make restaurants more desirable to the health conscious diner.

Despite the positive comments from diners, we also identified perceived barriers. More than half reported that they thought that current healthy options were too expensive and slightly less than half thought that healthy items would not taste good or be satiating. This indicates strong room for improvement in how chefs and operators develop their menus. Although this can be difficult for Chefs and operators to achieve, changes do not need to involve entire renovations of their menu's, but can involve making small changes to their top dishes so they become slightly healthier. Chefs and operators can also improve the descriptions of their healthier dishes, beyond simply stating "healthy option", so they sound more appealing to diners and inspire them to choose that dish. This is vital, as diners eat with all of their senses and the words they see to describe menus can have a big impact on choice. Descriptive names not only improve the perception of the how the item will taste and the enjoyment of the food, but it also improves the sales and leaves a favorable impression of the restaurant in the diner's mind (Wansink, van Ittersum, & Painter, 2005).

One striking finding from the current study was the strong inter-country differences in terms of desires and barriers to choosing healthy meals in restaurants. For example, in Brazil and China the frequency of choosing health meals was much higher than in Germany and Russia. Moreover, in Indonesia and China the current level of satisfaction was quite high compared with the UK and Germany, where complete satisfaction with current offerings was quite low. These inter-country differences could result from a myriad of influences, for example they could represent a large difference in current meal offerings in the countries, the traditional eating patterns and the perception of foods as healthy or otherwise. Irrespective of the cause of these differences the implication is clear that interventions and programs to improve the healthiness of restaurant meals need to be tailored to local markets and needs to ensure they are appealing to local diners.

This research establishes the needs and barriers of diners, however further research will also help to enhance our knowledge and the ability to provide targeted services. For example, we now know what type of small changes diners want to see, but it is also of interest to determine which type of dishes diners are prepared to see changes on. It is also of use to know more about the best approach to communicating small changes to diners. In the past, many chefs and food service operators have taken a "stealth health" approach. That is, the dish might be healthier but there is no desire to call out the healthy features for fear diners will reject it. However, there is some indication that health and nutrition is a growing trend in the restaurant industry. The National Restaurant Association's What's Hot 2015 Chef Survey reveals that health and nutrition are "hot" with diners, especially healthier children's meals, vegetarian dishes and sustainable sourcing (National Restaurant Association, 2015). Transparency on the menu might be the right complement to a stealth approach to health and nutrition in the food service industry. Either way, menu items with appealing descriptions that show no taste compromise are of interest to the vast majority of diners.

The current study had several strengths, such as a comprehensive questionnaire and an international population. However, there are also some limitations which should be taken into account when interpreting the findings. First, the current study was completed using an online panel which was restricted to those that ate out at least once per week, which may have led to an underrepresentation of people from a lower socioeconomic status in the sample. This could limit the generalisability of current findings to current restaurant eaters and those that are from a slightly higher socioeconomic status. Second, the definition of 'healthy' in the current study was deliberately left undefined to study respondents, as (a) 'healthiness' of meals is hard for diners to gauge, and (b) we wanted to understand without educating diners what they currently thought of restaurant meals in terms of their own perception of healthiness. While this gives interesting insights into current perceptions of diners it could mask some findings, for example in cases where diners inaccurately perceive some restaurant meals as healthy. Finally, some of the terms used in the questionnaire, such as the meal times and occasions, may be less appropriate for some cultures which may affect the interpretation of the results for some countries, however they were chosen to provide a comparison point across countries.

Bringing healthier meals to restaurants that are appealing to diners could have an impact on overweight and obesity. However, improving health in these settings is not just the responsibility of chefs and operators, there are also many other key players that are needed to work together to ensure success. While chefs and operators can make a kick start to leverage efforts, everyone associated with the food service industry needs to participate, all the way from governments to diners. All of these key players are involved in either indirect changes (e.g., new regulations) or direct changes (e.g., enabling individual behavior change through education and awareness programmes). Given the large number of chefs and operators and the large number of meals catered for, these small, but vital changes can really add up to make a big difference towards the global overweight and obesity epidemic.

5. CONCLUSIONS

In summary, the majority of diners were not completely satisfied with current healthy options, highlighting the need for improvement. Diners wanted to see small, healthy changes and responded clearly about which items they wanted to see included by chefs and operators on menus to make them healthier, if they are not already present on menus. Overall, healthy options were often linked to lower taste and satiation and higher price, as such menus reporting healthy dishes need to be altered to ensure these barriers are overcome. Diners had clear preferences for when they wanted healthier meals. Chefs and operators from all sized operations (small independent to large multi-unit food service) can use this information to know when to highlight healthier meals, and when more work is needed to sell these dishes. Differences in the needs and wants of diners per socio-demographic and personal factors were present indicating the need to tailor menus and the manner of selling meals locally.

CONFLICT OF INTEREST

R.S.N., R.vd.M., A.B. and L.C. are employees of Unilever R&D. Unilever produces foods which are marketed to fit in a healthy diet and lifestyle and is the owner of the Unilever Food Solutions Service. "Seductive Nutrition™" is a trademarked term, owned by Unilever Food Solutions. CR was sponsored for her time given to this research by a grant from Unilever and is on the speaker's bureau for the Gatorade Sports Science Institute.

AUTHOR INVOLVEMENT AND DISCLOSURES

All authors have been involved in the design of the study, drafting or revising of the manuscript for intellectual content and approved the final version to be published. Preliminary data from this research in a basic form was released as a non-academic PR material report "Seductive Nutrition" in a Unilever series entitled the World Menu Reports. A copy of this booklet can be downloaded at: http://www.ufs.com/company/media-center/world-menu-report/world-menu-report-3.

ACKNOWLEDGEMENTS

We would like to thank the participants for their time and effort. We would also like to acknowledge the efforts of everyone involved in the data collection process and development of the Seductive Nutrition™ service, in particular Mathilde Terink, Bas van Waveren, Kees van Erp, Lynn Gately and Eva Skoog.

REFERENCES

1. American Heart Association. (2013). Heart disease and stroke statistics-2013 update: A report from the American heart association. Circulation, 127,143–152.

2. Data Monitor. (2013). Data monitor consumer trends. Data Monitor.

3. Gilden Tsai, A., Williamson, D. F., & Glick, H. A. (2011). Direct medical cost of overweight and obesity in the United States: A quantitative systematic review. Obesity Reviews, 12, 50–61.

4. Heshmat, S. (2011). Eating behavior and obesity: Behavioral economics strategies for health professionals. New York: Springer Publishing Company.

5. Hill, J. O. (2009). Can a small-changes approach help address the obesity epidemic? A report of the joint task force of the American Society for Nutrition, Institute of Food Technologists, and Information Food Information Council. American Journal of Clinical Nutrition, 89, 477–484.

6. McCrory, M. A., Fuss, P. J., Hays, N. P., Vinken, A. G., Greenenberg, A. G., & Roberst, S. B. (1999). Overeating in America: Association between restaurant food consumption and body fatness in healthy adult men and women ages 19 to 80. Obesity Research, 7, 564–571.

7. National Restaurant Association. (2015). What's hot 2015 chef survey. Available at http://www.restaurant.org/Downloads/PDFs/News-Research/WhatsHot2015- Results.pdf http://www.restaurant.org/News-Research/Research/What-s-Hot [Accessed 19.01.15].

8. Nicklas, T. A., Drewnowski, A., & O'Neil (2014). The nutrient density approach to healthy eating: Challenges and opportunities. Public Health Nutrition, 17, 2626–2636.

9. Papies, E. K., & Veling, H. (2013). Healthy dining: Subtle diet reminders at the point of purchase increase low-calorie food choices among both chronic and current dieters. Appetite, 61, 1–7.

10. Thaler, R. H., & Sunstein, C. R. (2013). Nudge: Improving decisions about health, wealth and happiness. New Haven, CT: Yale University Press.

11. Todd, J. E., Mancino, L., & Lin, B. H. (2010). The impact of food away from home on adult diet quality (Rep. No. 90). United States Department of Agriculture.

12. Wansink, B., Just, D. R., & Payne, C. R. (2009). Mindless eating and health heuristics for the irrational. American Economic Reviews, 99, 165–169.

13. Wansink, B., van Ittersum, K., & Painter, J. E. (2005). How descriptive food names bias sensory perceptions in restaurants. Food Quality and Preference, 16, 393–400.

14. World Health Organisation. (2014). Obesity and Overweight. WHO Fact Sheet series, No. 311.

INDEX

Q

R

S

T

V

W